More praise for *The Art of Business Value*

"With an engaging writing style, Mark Schwartz builds a case for developing an organizational culture for . . . IT and other stakeholders to collectively work to define and deliver business value. If you care about how IT can effectively support improved outcomes for your organization, you should read this book."

Richard A. Spires, CEO of Learning Tree International, former CIO of US Department of Homeland Security, and former Deputy Commissioner and CIO of the IRS

"Mark Schwartz masterfully deconstructs the overloaded and misunderstood concepts of 'business value,' 'product owner,' and other nebulous value terms. He then introduces us to novel approaches that directly address many of the shortfalls of more traditional systems—finally helping to plug a gaping hole in much of the Agile literature."

Max Keeler, Chief Projects Officer of the Motley Fool

"This book brings together Agile, DevOps and Business Value in a coherent, direct yet humble manner with insights that can only come from an experienced practitioner. 'At the heart of DevOps is a machine—a Continuous Delivery pipeline . . . The pipeline is an automated bureaucracy . . . it applies its rules in a rigorous, unemotional way' – priceless!"

Tapabrata Pal, Ph.D., Director & Platform Engineering Fellow, Enterprise Application Platforms

"Schwartz calls into question all previously known techniques for assigning business value and provides a sane, modern approach to evaluating priority. Bridging the gap between Agile, DevOps, Lean Startup, Continuous Delivery and the business, he makes his recommendations approachable, couching them as experiments . . . a must have book for anybody interested in a better way to align their IT department with the business."

Mike McGarr, Engineering Manager, Netflix

THE ART OF BUSINESS VALUE

THE **ART** OF
BUSINESS VALUE

Mark Schwartz

IT Revolution
Portland, Oregon

IT Revolution
Portland, Oregon
info@itrevolution.net

For quantity purchases by corporations,
associations, and others, please contact the
publisher at orders@itrevolution.net.

Cover and book design: Stauber Design Studio
Illustrations: Lauren Simkin Berke

To those who are deploying code.
We can talk all we want about business value,
but they are the ones creating it.

CONTENTS

A mystery solved, in seven chapters.

1   CHAPTER 1: **THE PROBLEM**

Agile and Lean practices are all about maximizing the business value we deliver. Agile books tell us that business value is important, but none of them seem to tell us what business value is, though they do scatter clues here and there. In this chapter, however, we examine the clues and find that they come to nothing. Business value is a mystery at the core of Agile practice.[1]

17   CHAPTER 2: **THE MEANING**

Our first step in solving the mystery is to consult the experts. Unfortunately, the experts just deepen the mystery for us.[2] It turns out that we need to cast a wider net to catch our elusive target.

37   CHAPTER 3: **THE CULTURE**

Good detective work involves observing people. We find that there seems to be a connection between organizational culture and business value. But to understand it, we have to immerse ourselves in corporate culture, rather than reject it or stand apart from it.[3]

51   CHAPTER 4: **THE RULES**

What could bureaucracy possibly have to do with business value? A lot, perhaps. Warning: this chapter contains graphic depictions of bureaucracy being applied to agility. You may come away wanting to produce burndown charts in triplicate.

1   cf. Jean-Paul Sartre's claim that "nothingness lies coiled at the heart of being—like a worm," in *Being and Nothingness*.

2   cf. Dante, *The Inferno*. It gets worse and worse the deeper you look.

3   Alienation has been covered thoroughly—some might say excessively—in *The Stranger* by Camus and "Bartleby, the Scrivener" by Melville.

FOREWORD

The study of business value seems obvious at first—after all, over the course of our careers, we've all seen which activities create value and which ones waste everyone's time. But what if our grasp of what business value really is, is not quite right to start with? In his new book, *The Art of Business Value*, the indomitable Mark Schwartz shows us that understanding business value is not as simple as it seems.

For me this book is reminiscent of a book that I love and have read several times over, *Zen and the Art of Motorcycle Maintenance: An Inquiry Into Values*, especially the passage in which Robert Pirsig tries to define "quality." Similarly, Mr. Schwartz points to many things we already know and have learned—but then he makes us pause and reevaluate as he shows us how many things we don't know, or didn't adequately question.

A particularly unsettling moment occurred for me when Mr. Schwartz points out that business value is often incorrectly conflated with either customer value or user value—in that moment, the book's goal of more precisely defining business value suddenly took on much greater significance and urgency. If we run our organizations to create value, are we correctly defining what types of value we strive to create? How do we measure it? And, by the way, whose job is it to define value, anyway?

Irreverent and whimsical, *The Art of Business Value* challenges conventional thinking and questions many of the deeply held beliefs of the Agile community. It forces us to examine carefully the concepts and definitions we thought we understood, which, in the end, allows us to define more precisely what business value is—so that we can create more of it.

Using a wide-ranging, educational, and scholarly exploration that covers nearly a century worth of organization design, business principles, and software delivery, *The Art of Business Value* offers a startling and incredibly rewarding journey for the reader.

Fearless and entertaining, this book is ultimately a quest to examine the concept of business value—a concept that we so often take for granted. It provides tools on how to better understand it and, more importantly, create it.

I found reading this book to be immensely satisfying, and I felt more informed and much smarter after reading it. I genuinely hope that you have as much fun and learn as much as I did as you read *The Art of Business Value*.

Gene Kim
Portland, Oregon
January 2016

Don't for heaven's sake, be afraid of talking nonsense! But you must pay attention to your nonsense. ... Never stay up on the barren heights of cleverness, but come down into the green valleys of silliness.

Ludwig Wittgenstein, *Culture and Value*

PREFACE

For the last few years, I have been struggling to bring Agile and DevOps practices into a large federal government agency—seemingly the most inhospitable of environments. As a newcomer to government, my first reaction was one of amazement: How was it possible to throw so many obstacles in the way of good practice? How was it possible to inject so much waste into processes that were otherwise headed in the right direction? What was especially striking, though, was that the obstacles and the waste were being injected by some of the most intelligent people I'd ever worked with, and certainly the most committed and well-meaning. It had the feeling of a paradox, set up by some very clever philosopher with two hundred-some-odd years to get the confounding details just right. As an ex-philosophy graduate student, I recognized that there was just one thing to do: approach it with a sense of humor and enjoy the elegance and aesthetic pleasure of an argument well delivered as I figured out its strange internal logic.

I haven't seen this problem just in Washington, DC. Before joining the government, I was the CIO for Intrax Cultural Exchange, a medium-sized, services-oriented, mission-driven company. The personalities of the founders were so strongly stamped on every interaction in the company that the organization seemed to have its own peculiar logic. Before that, I proudly played a role in the dot-com bust of the early 2000s as CEO of a small software startup and a general hanger-on in Silicon Valley. Delivering software product seemed to rely on yet a different kind of logic: Could you make sure your features rolled into a one-sentence pitch? Good talent and good ideas were going unfunded—economic waste—but there was a logic to it if you didn't have to take it too seriously.

In the classic literature of Agile software development, teams produce business value. In the Scrum model, there's someone called a product owner who figures out what is valuable by applying some ROI standard; there are customers waiting to be pleased; self-organizing teams stand ready to shear away the stuff that they somehow know is waste. I'm sometimes amused at how simplistic the notion of business value is. The question the authors take on is how to create lots of this value stuff, whatever it is, as soon as possible—and cause the cultural change that will make it possible. But in the actual situations I have faced, organizations seemed to have strange ideas about what value looked like.

Here's what I realized: the strange notions of value expressed by each of these organizations actually made sense in their contexts. The odd behaviors of the government agency, the closely held company, and the startups competing for venture funding were entirely rational and appropriate to their circumstances. Activities that seemed wasteful were not always so; priorities flowed from what was *really* important, not from some universal standard. The *meaning* of business value—not just the features that would realize it—was different from organization to organization.

What if—humor me for a moment—some of that waste that the government injects into its processes actually adds a kind of business value? What if business value in the startup community means raising capital at higher and higher pre-money valuations, and generating profits is only a distant second as a business goal? It didn't make much difference to us in a Waterfall world—we cared about schedule and cost milestones. But in Agile practice, we only care about the delivery of business value. Which means we care about . . . what?

That was the train of thought that led to this book. The more I explored the topic, the more critical it seemed to become. It seemed to have implications in how Agile teams fit into the enterprise, how we measure their success, how we go about causing cultural change, how we think about the IT function in a company, how we deal with compliance and bureaucracy, and how we choose and work with a product owner or on-site customer. It was about whether the way we practice Agility aligns with the philosophy behind it. It was about how I should do my job as a CIO.

Humor often requires that we accept the bizarre logic of an unfamiliar world. Consider this exchange from *Through the Looking Glass, and What Alice Found There*:

"I see nobody on the road," said Alice.

"I only wish *I* had such eyes," the King remarked in a fretful tone. "To be able to see Nobody! And at that distance too!"

Once you accept that a little girl has gone through a mirror or down a rabbit hole and is having conversations with a Cheshire Cat and a childish king, who seem to be logicians and always find ways to twist her words around, the exchange makes a lot of sense. All I'm saying is that business value is sort of that way, too.

I feel some urgency about this whole thing.

Have you ever been to one of those big Agile conferences and seen all the people wandering around, trying to decide which of the many sessions to attend? "The Armadillo Model: *Dasypus hybridus* and the Snuffling Anti-Pattern,"[1] "50 Shades of Agile,"[2] "Lessons from Coaching a Cult of Dancing Schizophrenics to Conduct Effective Retrospectives." There are a lot of ideas out there. Sometimes it helps to think about what ties them together—that's right, this business value thing. The *raison de backlog*.

Our abstraction is leaking.[3] I don't know about you, but I've found it pretty hard to locate a good product owner. We know we are creating features, but are we creating business value? Sure we are: the product owner says so, and she's from *the business*. Are you frustrated trying to explain to her why spending time to reduce technical debt is more important than adding the mimsy borogoves feature to the Jabberwock, which Marketing says will have an ROI of 321.25 percent? "But refactoring the FixAllYourProblems class to use the Classy Recursion pattern is worth 7.2 Value Points!" I don't have exact stats for you on how many of us have tried unsuccessfully to have this business value conversation, but it's a lot. Anyway, 97.5 percent of readers believe that statistics in books are mostly made up.[4]

In the meantime, the cutting edge of Agile practice today—DevOps and Continuous Delivery—seems to be moving us toward smaller and smaller batch sizes of requirements, perhaps approaching single-piece flow. Or maybe even smaller—some organizations seem to be deploying change sets so small that they're just fractions of features. It starts to feel like a calculus problem—what is the limit of risk as requirements batch sizes approach zero? Our business case is vanishing into infinitesimals with a smile on its face, like the Cheshire Cat.

Then there's the CIO, the executive who's in charge of making sure that IT projects have business value. Or was that the product owner? Well, at least the CIO is in charge of delivering solutions, then. Or was that the Agile team? Have you noticed all the books telling CIOs how to be better CIOs? I have, because I'm a CIO. Mostly they say that the CIO should grab a "seat at the table" (that's the executive table, where the grown-ups sit). Perhaps this makes sense, because the CIO will need a place to sit while the Agile teams are out creating business value. What's a CIO to do in a world where their teams are off plotting business value with the business?

One more thing to point out. As those teams are off deploying sashimi slices of value, where do we think those slices wind up? They immediately become legacy slices of sashimi, and I promise you that that is not a good thing. The boundaries of our systems are blurring with loose coupling and microservices. So ultimately they join that giant agglomeration of IT capabilities that we sometimes call the Enterprise Architecture (EA). I don't mean EA in the sense of a bureaucracy of Visio abusers; I'm talking about that asset, the abstracted total of IT capabilities that allows the business to operate—software, infrastructure, and all that. That giant hairball of stuff that the CIO oversees, that keeps getting new features stuck to it—duct tape, rubber cement, chewing gum, etc. (And a few bits of mixed metaphor, too.)

The hairball has economic value, clearly, since it enables the business. We are just turning the corner on how we think of risk and value in our IT *projects*—should we also be thinking in terms of the value of our hairball? How are we going to care for the hairball, keep it rolling in the right direction, and pretty it up? (Let me introduce a technical term here: "Ick.")

What I'm saying is that business value is a problem.

So this book is a bit of a meditation on business value and why it matters to us. Or maybe it is more of a detective story. Business value is out there somewhere, even as our deployments become vanishingly small, and we're going to track it down. We'll interrogate the usual suspects, round up some experts, recruit some informants, test out some theories, and, in the end, track it down. Then we'll get it to work for us, if the government says it's authorized to work.

Some things will occur to us as we follow the footprints. Every decision we make in a software development project is ultimately a decision about business value. Feature trade-offs are decisions about business value. Risk management is about business value. Communication with the enterprise is about business value. Developer morale is about business value—it can affect the company's costs in hiring and retaining developers, and it can affect the inclination of developers to innovate new value-creating solutions. Agile thinking is explicitly about business value: instead of delivering to schedule milestones, we deliver simply in the way that maximizes business value.

But we'll also see that "business value" is often implicit, or at least rarely explicit enough for someone to act upon. We'll recognize that while learning organizations are important to us, the value of learning is often unstated and the learnings themselves are rarely explicitly valued. We'll pause to consider what the job of IT leadership might be, especially given that more and more responsibilities are being pushed down to the teams. Ultimately, we'll arrive at an idea of business value that I think is consistent with today's thinking about organizations. Then we'll look at how that understanding should influence the way we practice agility. Perhaps we'll even talk about how to polish up that hairball.

But for now I'll just follow the advice of Alice's King of Hearts: "Begin at the beginning," the King said gravely, "and go on till you come to the end; then stop."

WHO THIS BOOK IS FOR

I wrote this book for the Agile practitioner community and for the wider IT community.

If you are an Agile practitioner—a developer, tester, or Scrum master—then you face decisions of business value every day. You need to be able to speak the language of business value to communicate with the organization at large. Your success or failure depends on your ability to create business value. (Are you more interested in business value now?)

If you are an IT specialist in operations, infrastructure, security, or just about anything else technical, DevOps is making you part of the Agile delivery effort. While your role was always about business value, it is now more explicitly so. You will feel more a part of the team if you are all aligned behind a common understanding of what it means to deliver value.

If you are a product owner or a representative of the business explicitly charged with delivering business value and responsible for prioritizing features based on business value, I strongly urge you to follow the discussion in this book. You have a hard job. In fact, you are being asked to do something impossible. Let me explain how to turn it into something possible.

If you are an Agile coach, a thought leader, a pundit, or a writer on Agility, I hope that I am saying things that you already feel in your bones. There has been a gap in our literature on the subject of business value, and I hope this book will address it. But even more, I hope that you will find ways to take ideas from this book and turn them into practice: my goal here is more to provoke and discuss, rather than to prescribe.

If you are a CIO, you are undoubtedly confused by the Agile literature, which has forgotten to mention you. This book is for you.

If you are an investor, you need to sound knowledgeable. It's all here. (And invest in my next startup idea, which will not only be clear on what business value is, but deliver lots and lots of it!)

If you are in management and love to make up rules (*requisitiphilia*), go straight to chapter 4.

If you are anyone else, this book wasn't directly written for you, but if you are curious, go ahead and pick it up. I will try to entertain you and give you some insight into how IT practitioners and Agile delivery teams think. You may have to look up a few of the terms I use, but you will have no trouble finding explanations online or in other books.

---

1 While I was researching this, I learned that there are actually many different types of armadillos, including the screaming hairy armadillo and the greater fairy armadillo. That kind of learning makes all that time in the library worthwhile.

2 This one's real. My colleague Josh Seckel presented it at Agile 2015.

3 Joel Spolsky, "The Law of Leaky Abstractions," http://www.joelonsoftware.com/articles/LeakyAbstractions.html. I just heard about this and knew I'd have to get it into the book somehow.

4 Mark Schwartz, *The Art of Business Value* (Portland, OR: IT Revolution, 2016), xvi.

The right understanding of any matter and a misunderstanding of the same matter do not wholly exclude each other.

Franz Kafka, *The Trial*

For I found myself embarrassed with so many doubts and errors that it seemed to me that the effort to instruct myself had no effect other than the increasing discovery of my own ignorance.

René Descartes, *Discourse On Method*

THE PROBLEM

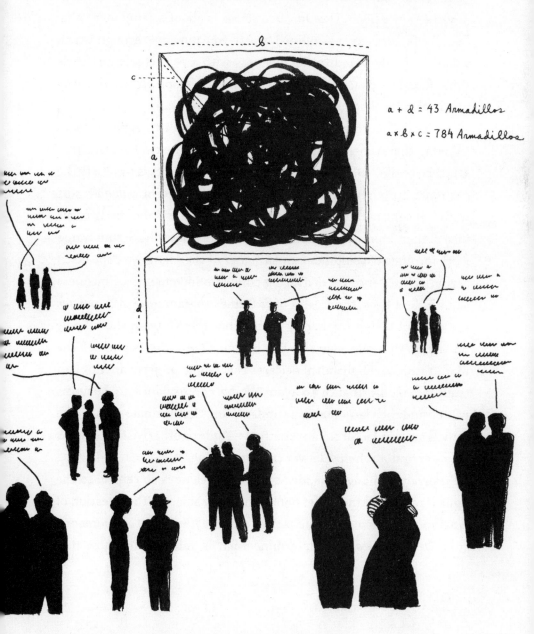

$$a + d = 43 \text{ Armadillos}$$
$$a \times b \times c = 784 \text{ Armadillos}$$

A core principle of Agile and Lean theory is that software development projects should seek to maximize business value. Projects should be judged not on their adherence to cost and schedule milestones, but on their delivery of value to the enterprise. Value should be delivered as quickly as possible—in small increments—and features should be prioritized based on the amount of value they deliver. DevOps, in a sense, is about setting up a value delivery factory—a streamlined, waste-free pipeline through which value can be delivered to the business with a predictably fast cycle time. Rapid feedback from production to development then allows us to optimize that value delivery machine.

The idea of business value was central enough to Agile ways of thinking that it merited a place at the head of the twelve principles attached to the Agile Manifesto: "Our highest priority is to satisfy the customer through early and continuous delivery of *valuable* software."[1] Several of the signers of the Manifesto later elaborated on this idea in their books. Ken Schwaber, the cocreator of the Scrum framework for Agile development and a signer of the Manifesto, speaks of Scrum's "insistence on delivering complete increments of business value."[2] Kent Beck, the creator of Extreme Programming (XP), pushes the concept a step further by saying that the XP team should *only* do things that add value to the business.[3] Another signer of the Manifesto, Jim Highsmith, declares that "Agile projects are not controlled by conformance to plan but by conformance to business value"[4] and then later makes a similar claim: "In the final analysis, the critical success factor for any method—Agile or otherwise—remains whether or not it helps deliver customer value."[5]

Strangely, although the idea of business value is so central to the Agile way of thinking, most books on agility sidestep the question of what exactly business value is. Instead, they assume that someone from "*the business*" will determine what is valuable and how that source of value should be translated into features and priorities. In Scrum practice, this person is the product owner.[6] The product owner is sometimes described as the visionary who steers the product and sometimes as the steward of business value decisions: the person who maximizes business value by making appropriate prioritization and scope decisions. In either case, the product owner provides the business value context to the team.

Also interesting is the vacillation, shown in the quotes above, about whether the goal of Agile development is the delivery of *business* value or *customer* value. Highsmith, you will notice, switches from one to the other in the course of twenty-six pages. The first principle in the Agile Manifesto is ambiguous—it speaks of satisfying the *customer* by providing *value*. Is business value the same as customer value? Many of the influential Agile thinkers and writers come from product-focused software companies, so it is natural they would think in terms of customers and their needs. Product-focused companies earn their revenues by delivering value to customers, it is true—but is *that* value the same as what we mean by business value?

The word *customer* is ambiguous in this context. If we take it to mean the buyer or user of a company's commercial software product, then the answer is no. While customers might want or value a particular feature, the business might not value giving it to them, for reasons of cost, maintainability, or consistency with the company's brand or competitive positioning. Features that deliver customer value do not necessarily lead to increased revenues, or they can be more expensive to develop than the revenue they drive. On the other hand, we can easily imagine software features that are valuable to the business even if they are *not* directly valuable to the business's customers: for instance, business intelligence reports, accounting functions, and procurement systems for supplies. Or consider a business whose strategy is to deliberately lose the 10 percent of its customers that are the least profitable—the ones who cost too much to serve and provide little revenue in return. In this case, adding business value may mean deliberately destroying customer value.

Of course, we do not have to take such a literal interpretation of the word *customer*. Perhaps the writers mean to include all of the users of the software, even if they are internal to the company. It seems obvious that a feature cannot be valuable unless it adds value for the person who is using it. That is why Agile approaches emphasize working directly with end users and continually soliciting their feedback. But this broader concept of *user* value still does not quite capture what we mean by business value. In the case of a transformational business initiative, for example, management wants to create fundamental change in the organization's processes, but individual users in the organization may not share that vision or may not be expert in interpreting and applying it. They might have smaller, more "local" priorities than the big-picture transformation that management has in mind. By trying to maximize what users consider to be valuable, the Agile team might simply be perpetuating old ways of doing things rather than contributing to a transformation that the business values.

In speaking to users about what they need from a piece of software, I've found a common pattern: they believe that processes that take them a number of steps should be automated to make their jobs easier. That can be very valuable to the enterprise—or not. The user might not realize that automation might lock in a process that is likely to change and might not factor in the costs of maintaining the software as that business process changes, for example. There can be many reasons why *business* objectives differ from *user* objectives.

Perhaps the authors mean to suggest that the business *as a whole* is the customer of the Agile development team. IT organizations have often been thought of as customer service organizations whose goal is to satisfy the needs of internal customers. Certainly contract software development shops think in terms of satisfying a business that is their customer. If the organization as a whole is the customer of the Agile team, then the alignment between customer value and business value is exact. But is this model of *the business* as the customer the appropriate model for an Agile organization? I'm not so sure, and chapter 5 will explain why.

I would like to suggest that the conflation of business value, customer value, and user value is outdated and well out of step with current Agile practice. As with requirements in general, we can no longer think of business value as something known and understood in its entirety before the team begins its work. More importantly, we cannot think of business value as something determined outside the team by something called *the business* and then simply presented or "tossed over the wall" to the team in the form of user stories, prioritization, and feedback on product as it is produced. The responsibility for understanding and interpreting business value cannot be placed solely in the hands of a product owner. And if the success of an Agile project is to be determined by the value it delivers, then we have to think of that value in terms of *outcomes*, not completed stories, and measure it as such. Releasing code is not the same thing as delivering business value; to know that we have delivered business value, we must both understand what business value is and be attentive to outcomes.

This might sound like an academic exercise: business value probably sounds about as interesting to Agile practitioners as bookkeeping and accounting—things that MBAs, people inclined to that sort of stuff, study in business school. I assure you that this is a mistake. A good understanding of business value is critical to Agile practice, and I will demonstrate that the question of business value becomes stranger and more revealing the more one examines it. It is critical, for example, in distinguishing between waste and value-adding work. I will try to show that many of the difficulties we routinely face in adopting and improving software development practices in an organization can be traced to business value and its interpretation.

We must admit that there is something tautological when we say that the goal of Agile software development is to deliver business value. Business value, intuitively, is whatever the business values, and the goal of every person and function in the business is to do what the business values. To say that we want to deliver business value is to say nothing much except that we want to do the right thing, do lots of it, and do it quickly. But this does not help us understand how to select and prioritize features.

In his 2011 blog post "The Elephants in the Agile Room,"[7] Philippe Kruchten tells of the signers of the Agile Manifesto returning to Snowbird, where the Manifesto was drafted, ten years later to discuss the difficulties they saw in the way Agile had been adopted. The thirteenth "elephant in the room," according to Kruchten, is that business value is "mentioned everywhere, but not clearly defined, or pushed onto others to resolve." Perhaps this is also related to the twelfth elephant they listed: "Abdicating responsibility for product success (to others, e.g., product owners)."

The question of business value is the question of purpose, motivation, mission, and direction. It is a question of value and values. If we build an elegant Continuous Delivery pipeline that harmonizes Development and Operations and continually checks its own health by feeding back from production, we have accomplished . . . what, exactly? It depends on what business needs we push through that pipeline, and what business value results from that. DevOps is form without content until we address the question of what goes in to the pipeline and what happens when product emerges at the other end.

It is comforting to think that business value is something well understood by *the business* and encapsulated in an objective metric. To the extent that the Agile literature talks about business value, it often puts it in the context of something called Return on Investment, or ROI. In the Scrum framework, the product owner is often seen as delivering business value by maximizing ROI. According to Mike Cohn, one of the clearest and most prolific writers on Agile practice, "the product owner is responsible for making sure the project earns a good return on the investment made in it."[8] Returning to Ken Schwaber's book, we find that "the product owner's focus is on return on investment (ROI). The Product Backlog provides the product owner with a powerful tool for directing the project, sprint by sprint, to provide the greatest value and ROI to the organization."[9] It is interesting that he says "the greatest value *and* ROI," implying that those are two different things, though elsewhere he seems to use the terms interchangeably. A group of Agile and Lean thinkers worked together in 2005 to formulate a Declaration of Interdependence, which includes as a foundational principle that "*we increase return on investment* by making continuous flow of value our focus."[10] The fact that ROI has a name, an acronym, and sometimes a formula makes it sound reassuringly precise. We are probably aware that the product owner is not actually calculating an ROI metric for each user story, but we feel that the standard is at least approximately being applied.

It is curious, once again, that ROI is not defined or explained, though we are told that Agile practice is all about maximizing it. Schwaber leaves us up in the air with a comment that "the product owner is responsible for the ROI of the project, which *usually* means that the product owner chooses to develop functionality that solves critical business problems."[11] *Usually*? What else is ROI, then?

Let's take a close look at a passage from two excellent Agile thinkers, Craig Larman and Bas Vodde:

The product owner is responsible for maximizing return on investment (ROI) . . .
The product owner has profit-and-loss responsibility for the product,
assuming it is a commercial product. In the case of an internal application,
the product owner is not responsible for ROI in the sense of a commercial
product (that will generate revenue), but they are still responsible
for maximizing ROI in the sense of choosing—each sprint—the highest-
business-value, lowest-cost items.[12]

It feels to me like these authors are struggling. Is it profit and loss the product owner is responsible for, or ROI—or are those the same thing? Does ROI mean something different for internal products than for external products? Is ROI the same thing as "highest-business-value, lowest-cost"? Are we going in circles, defining business value in terms of ROI, which is then defined in terms of business value?

A page later the authors get themselves into deeper trouble. Explaining the practices the product owner must use, they say that "the product owner prioritizes the backlog . . . to maximize ROI (choosing items of high value with low effort) or secondarily, to reduce some major risk."[13] Hang on! Is risk part of ROI, or is it a whole separate thing the product owner has to worry about?

I've chosen this passage from Larman and Vodde—two authors I respect—to show what I think is the typical vagueness and imprecision with which questions of business value are addressed in Agile literature, even while the authors agree that business value is the most important thing to focus on.

———————

Well then, is ROI the same thing as business value? Does maximizing ROI maximize business value? Are we even sure we understand what ROI is in the first place?

We probably don't. In the financial world, ROI is actually not well defined. Everyone agrees that it is calculated by dividing the return from an investment by the cost of the investment. The difficulty is that the "return" in the equation can be pretty much anything. Return is the good stuff that we get by investing, whatever that might be. The most commonly used numerator for ROI is profit, or earnings (the two terms are equivalent). But making investment choices based on a function of profit, as we will see, can lead to poor decisions.

Why not simply use sales, or revenues? Because we are building a set of features that customers value, shouldn't we measure value by the sales that result? For one thing, focusing only on revenue would ignore any costs that the new features bring to the business. For example, will the new features increase our helpdesk support costs? Do the new features increase our sales of a physical product in such a way that we need to stock more inventory? If so, then revenue only gives us a piece of the value picture. And if the features we are developing are only used by employees internal to the company, perhaps to decrease costs, then revenue is not even relevant.

So profit, defined as revenue minus expenses, is a better measure. Of course, when the product owner is looking at the value of a feature, the total profitability of the company is not what is important, just the *incremental* profits that will result from the feature. And what time period's profits does she care about? Typically ROI uses the average profits for a given number of years into the future. Of course, when the product owner is prioritizing features in a backlog, she does not actually know what increases in profit will result from each feature; she only has projections to work from. She doesn't really even know how much the cost of the investment will be—she has only the developers' estimates of effort. So really we are defining ROI as used by the product owner as *projected* average annual increase in profits divided by *projected* investment cost. It turns out that this is the most common definition of ROI used by companies to make investment decisions.

Unfortunately, making decisions based on this ROI formula does not necessarily maximize business value.

The first problem is that profit does not consider the *timing* of the cash flows from sales and costs. As we will discuss in the next chapter, there is a time value of money that must be considered. ROI is a simple formula—that is its chief benefit—but it is misleading to simply consider short-term profits or to mix together short- and long-term profits.

Second, with ROI we are not considering the *risk* of the expected returns (or of the cost, for that matter). We are taking a point estimate of a projection, which discards important information about how certain the estimate is. There are different ways to factor in uncertainty: by using confidence ranges rather than point estimates, for example, or by reducing projected profits by a "risk factor." But simple ROI does neither of these things.

Thirdly, *profit* is based on financial accounting reports and is not intended for managerial decision-making. In financial accounting, cash flows are adjusted using an *accrual* method to give investors a picture of the company's health. Revenues and expenses are "recognized" in time periods that might be different from when the associated cash is received or disbursed. Depreciation and other non-cash expenses are factored in, as are increases in *working capital*, the temporary accumulation of inventory and credit given to customers. Accountants have considerable latitude in how to compute these numbers: for example, choosing depreciation methods and deciding whether to account for inventory using First-In-First-Out (FIFO) or Last-In-First-Out (LIFO) techniques. These decisions made by the accountants affect profit, but they do not affect the underlying economics of whether an investment is good or bad.

To decide whether an investment is worth making, companies compare the ROI to a *hurdle rate*, or minimum return that investors demand. But according to Alfred Rappaport in his book on how managers should maximize value for their shareholders, this makes no sense. "The essential problem with this approach," he says, "is that ROI is an accrual accounting return and is being compared to a cost of capital measure, which is an economic return demanded by investors."[14]

In fact, near-term profit is poorly correlated with the value delivered to shareholders of a company. In a classic textbook on how to measure the value of companies, Tom Copeland and his coauthors point out that changes in accounting technique that have *reduced* profits have often resulted in *higher* stock prices;[15] Rappaport, speaking of Earnings (i.e., profit) Per Share (EPS) reports that "numerous companies have sustained double-digit EPS growth while providing minimal or even *negative* returns to shareholders."[16]

Even if ROI were a good proxy for business value, it would not be very useful to product owners for prioritization decisions. In "The Problems with Estimating Business Value," Mike Cohn points out that it is difficult to assign value to individual stories, because the values of user stories are often intertwined. As examples, he asks what the values are of the left front wheel of a car or the doors and windows of a house.[17] None of these individually makes a difference in ROI, but presumably all are valuable. In a blog post entitled "How do you estimate the value of user stories? You don't," Pascal van Cauwenberghe questions the very idea of first writing stories and then estimating their ROI, since that can only result in a "vomit of user stories" that might or might not turn out to have value. Instead, one must "first determine what is valuable and then write user stories to deliver that value."[18] Dean Leffingwell, who has written extensively on Agile requirements, notes that prioritizing features through ROI is challenging because it involves making trade-offs between different types of value, and revenues generally cannot be allocated on a feature-by-feature basis.[19] So even if ROI were the right metric, it would be difficult to implement.

Perhaps Leffingwell is even understating the case when he says that quantifying returns is difficult. The practical aspects of projecting returns are daunting. For example, the product owner may project revenue increases for a particular feature, but what happens if a competitor copies that feature? Does the product owner really know how the new feature will affect the marginal profitability of the company? It can't be considered on its own, because it might affect other revenues and expenses of the company; that is, it might have side effects. Perhaps the new feature will cause increasing adoption of the product, but it will cannibalize other products that the company sells.

Remember that we've been speaking of ROI solely in the context of product companies. What if the software development effort is meant to serve users internal to the company? In this case, the impact on profitability may be even harder to ascertain. What is the impact on profitability of a dashboard that enables management to drill down on sales by region? There undoubtedly is a connection, but assessing it involves so many assumptions that the exercise is impractical. The new dashboard may occasionally help management spot and diagnose an issue that mid-level supervisors have not noticed, and that issue might lead someone to formulate a solution, and that solution might increase sales in a predictable way . . . but the product owner will be in a state of analysis paralysis before all of this gets worked out for prioritization.

I want to be careful here: although forecasting changes in profitability to make prioritization decisions seems impractical, I am not saying that it is impossible or that measuring actual changes in profitability after the feature is implemented is impractical. As Douglas Hubbard points out in *How to Measure Anything*, we can use statistical techniques to tease out how much of an increase in profits was due to a particular investment. We can also use measurements to reduce our uncertainty about a planned feature's impact on future cash flows. But I do not think this makes ROI a useful proxy for business value in prioritizing user stories.

Curiously, one of the things ROI does not take into consideration is agility itself. Part of the business value that software development can give us is the ability to respond to unknown future needs. We can build things in a way that gives us more options in the future or in a way that gives us validated learning about the environment we are in. In economic terms, we can say that software development efforts can give us "real options"—that is, options to invest more or to not invest in the future, depending on which way the market goes. This agility has true value to the organization, but it will not be accounted for in an ROI calculation. We will come back to this subject later.

We can fix some of the problems with ROI by using more sophisticated measures than incremental profit as the numerator of the equation. For example, we can look at incremental cash flows. We can even discount the cash flows based on timing and risk. But once we start moving in that direction, we start losing the value that ROI was intended to provide: simplicity in analyzing investment choices.

We will have to look elsewhere for the meaning of the elusive term *business value* that is the very core of our Agile practice.

The Problem: business value, critical but elusive, remains at large. Our first set of clues leads nowhere.

1   Various Authors, "Manifesto for Agile Development," February 11–13, 2001, http://agilemanifesto.org. The emphasis is mine.

2   Ken Schwaber, *Agile Project Management with Scrum* (Redmond, WA: Microsoft Press, 2009), Kindle loc. 207–208.

3   Kent Beck and Cynthia Andres, *Extreme Programming Explained: Embrace Change.* 2nd ed. (Boston: Addison-Wesley, 2005), 3.

4   Jim Highsmith, *Agile Software Development Ecosystems* (Upper Saddle River, NJ: Addison-Wesley, 2002), 32.

5   Ibid., 58.

6   Other Agile frameworks such as XP do not have an explicit product owner role. In XP, an onsite customer represents the business. But ultimately decisions are somehow being made about what will deliver business value. So when I refer to the product owner in these early chapters, please take it as referring to the person, people, or mechanism responsible for these decisions.

7   Both quotes in this paragraph are from Philippe Kruchten, "The Elephants in the Agile Room," February 13, 2011, http://philippe.kruchten.com/2011/02/13/the-elephants-in-the-agile-room/.

8   Mike Cohn, *Succeeding with Agile: Software Development Using Scrum* (Boston: Addison-Wesley, 2010), 125. Cohn has also written compellingly on Agile requirements and Agile project planning.

9   Schwaber, *Agile Project Management*, 18.

10 Jim Highsmith et al., "Declaration of Interdependence," February 17, 2005, http://pmdoi.org. Highsmith's cosigners on this document include many of the leading Agile thinkers—Mike Cohn among them.

11 Schwaber, *Agile Project Management*, Kindle loc. 1134. Emphasis is mine.

12 Craig Larman and Bas Vodde, *Scaling Lean and Agile Development: Thinking and Organizational Tools for Large-Scale Scrum* (Upper Saddle River, NJ: Addison-Wesley, 2009), 309.

13 Ibid., 310.

14 Alfred Rappaport, *Creating Shareholder Value: A Guide for Managers and Investors* (New York: The Free Press, 1998), 31.

15 Tom Copeland, Tim Coller, and Jack Murrain, *Valuation: Measuring and Managing the Value of Companies.* University ed. (New York: John Wiley and Sons, 1996), 85.

16 Rappaport, 5.

17 Mike Cohn, "The Problems with Estimating Business Value," *Mountain Goat Software*, September 30, 2010, http://www.mountaingoatsoftware.com/blog/the-problems-with-estimating-business-value.

18 Pascal van Cauwenberghe, "How do you estimate the business value of user stories? You don't." *Thinking for a Change*, December 30, 2009, http://blog.nayima.be/2009/12/30/how-do-you-estimate-the-business-value-of-user-stories/. The title says it all, doesn't it?

19 Dean Leffingwell, *Agile Software Requirements: Lean Requirements Practices for Teams, Programs, and the Enterprise* (Boston: Addison-Wesley, 2011), 261.

First learn the meaning of what you say, and then speak.

Epictetus, *The Discourses*

Make for thyself a definition or description of the thing which
is presented to thee, so as to see distinctly what kind of a thing
it is in its substance, in its nudity, in its complete entirety,
and tell thyself its proper name, and the names of the things
of which it has been compounded.

Marcus Aurelius, *Meditations*

2

THE MEANING

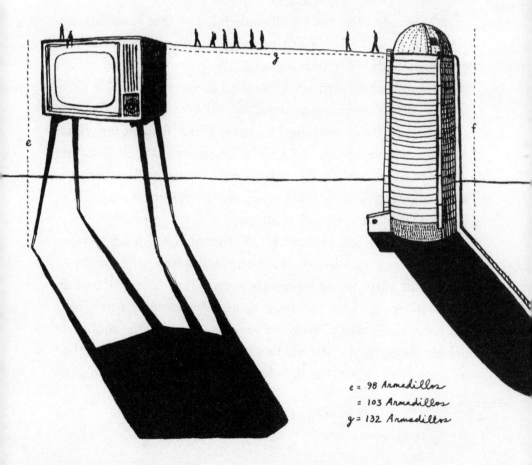

e = 98 Armadillos
= 103 Armadillos
g = 132 Armadillos

I f ROI is not the right measure of business value, then what is?

Ultimately, in a capitalist economy the duty of a corporation is to return value to its owners. Some writers have argued that the only way to deliver on this goal is to manage to it directly, rather than using proxy metrics like ROI. Such an approach is referred to as "Management by Value" or the "Shareholder Value Approach." According to this way of thinking, managers aim at making investments that maximize Market Value Added (MVA) or Shareholder Value Added (SVA). The technique is described well in books on valuation by Alfred Rappaport[1] and Tom Copeland.[2] To maximize MVA, these authors argue, managers must combine investment decisions with decisions about how to raise financing, signals from the stock market about its expectations, and decisions on when to return cash to investors as dividends instead of reinvesting it.

In Copeland's view, MVA is the only measure that takes into consideration *all* of the impacts that a project (or a feature set) will have on the company, side effects and all. MVA determines the company's long-term sustainability, since increases in shareholder value make more investors interested in investing in the company. It takes into consideration the future competitiveness of the company (or at least the market's perception of it) and is the metric that includes the interests of all other stakeholders, since equity holders have the "residual claim" on a company—they are the ones who are paid out last in a liquidation, after all creditors.

Copeland does note that outside the United States, business value is not always seen this way: in continental Europe and Japan, he points out, "intricate weightings are given to the interests of customers, suppliers, workers, the government, debt providers, equity holders, and society at large."[3] In his opinion, however, all of these interests are adequately represented in MVA. Rappaport goes as far as to define a "value ROI" metric: the shareholder value created divided by the cost of the investment.[4] Value ROI, he argues, should be used instead of accrual accounting ROI to make investment decisions.

But even managers who believe that MVA is the ultimate measure of business value don't, for practical reasons, use it for their everyday capital budgeting or investment decisions. It would be hard to imagine a product owner prioritizing features based on their projected impact on share price. Fortunately, there is a simpler metric that can often be used for making investment decisions consistent with MVA: Net Present Value, or NPV. NPV is a reasonably simple calculation that takes into consideration the cash flows that will result from an investment, their timing, the risk of the investment, and the opportunity cost of making an investment rather than returning money to shareholders as dividends. If made correctly, NPV-based decisions ultimately optimize MVA. Richard Brealey and Stewart Myers are the authors of a popular MBA textbook on corporate finance; in it, they say—perhaps surprisingly—that "the remarkable thing is that managers of firms can all be given one simple instruction: maximize net present value."[5]

Just what is NPV? Lest you think NPV is something just for *the business* to understand, I'm going to try to give you most of the value of an MBA program in just the next few paragraphs. Incidentally, a two-year MBA program itself has a rather low NPV. You might want to work it out as an exercise while I explain the concept. Make a few business assumptions and see if you can value the user story, "As an Agile practitioner, I would like to attend an MBA program so that I will understand what business value means." I'll give my answer later in this chapter.

I believe that the major lessons covered in an MBA program can be reduced to two principles:

1. There is a time value of money.
2. A business venture needs a sustainable competitive advantage.

Principle one says that a business should generate cash flows, preferably as soon as possible, and principle two says that in order to continue to generate cash flows, it needs a way to continue competing effectively in its market. You are probably not surprised that these things are true. What is interesting is how they apply to business decisions.

Suppose I propose an "investment" to you: you give me a $100 and I give you back $105. Are you interested in that investment?

A good answer is, "It depends." *When* do I give you back the $105? If I take your $100 and immediately give you back $105, it is certainly a good investment, and you should keep making it as long as I'm willing to offer it. The longer it will take me to give you back the $105, the less good the investment is, because you are without your $100 for a longer time. Let's say I propose that you give me the $100 now and I will give you the $105 in one year. Are you still interested in the investment?

Once again, a good answer is, "It depends." It depends largely on what other options you have for "investing" your $100. If you have another friend who says that he will turn your $100 into $110 in a year, then investing with me is a bad idea. If the only alternative you have is to put your money in a savings account that pays interest of 1 percent per year, my proposal sounds much better. So the value of an investment clearly depends on both how long it will take to pay off and what alternatives you have for investing the money.

This might be a bit unintuitive: you might not care how quickly you get your money back as long as you have plenty of other money available for your everyday needs. When we are talking about small amounts of money lent informally, it doesn't really matter to us if it takes time for the money to be returned, as long as we don't need it. But we *should* care if we have a viable alternative for earning interest on that money. A business is responsible to its shareholders and must make sure it earns a good return on any cash it has. For a business, the time value of money is critical.

Now suppose I say that the $105 I'm planning to give you back is not certain. I *think* I will be able to give you back $105, but the exact amount "depends on some factors," and I might not be able to give it back to you at all. Does this make the investment more valuable or less valuable? Less, of course. How much less? It depends on how risky the $105 is. Another way to look at it is that the higher the risk, the higher the return you should want to make up for the risk. If it's going to be risky, you might want more than $105 to make you comfortable with the investment. So the value of an investment depends on the timing of its payoff, the alternative investments available, *and* the risk associated with those payoffs.

That's the four-paragraph MBA.

You might be wondering about that second principle, the sustainable competitive advantage. Here's a way to think about it: when a business makes an investment, it is spending cash now in anticipation of a series of cash flows in the future. Let's say that we are developing a software product that will yield cash for us every year over the next five years. Notice that the value depends on our *projections* of cash flows into the future. How solid are those projections? Well, it helps if we are sure that our product can continue to stay ahead of the competitors. If you think about valuing a company as a whole—it is, after all, a sort of machine for producing cash flows—its value depends on its ability to *sustain* its cash flows. And that, of course, depends on whether it has a sustainable competitive advantage.

Okay, now to the value of getting that MBA. You will be investing two years of tuition and living expenses—let's say about $200,000.

You are also forgoing your opportunity of working over the next two years, which we will say is worth $300,000, since you are a well-paid software expert. So your total opportunity cost is $500,000. In return, you hope to earn more in the future. How valuable is that? It depends on how much more you hope to earn, how far in the future you will be earning that increased income, what alternatives you have for your money, and how risky that additional earning is. Let's say that instead of earning the $150,000 you are earning now, you believe that the MBA will allow you to earn $300,000 a year beginning five years after you graduate. So your increase in earnings will be $150,000 per year ($300,000–$150,000). You are thirty at graduation, thirty-five when your salary goes up, and you will work till you're sixty-five—so you will have thirty years at the higher salary.

Given that there is a time value of money, the higher salary you will get in the future is worth less than if you had it now. Your $500,000 cost, however, is all in the present. You want to know whether the $500,000 *now* is more or less than the $150,000 per year for twenty years starting six years in the future (simplifying and saying that your salary five years out is delivered in a bundle at the end of the year). We know it depends on the risk—how likely is it that you will hit that salary target—and how much you would earn if you invested money in an investment with a comparable risk. The risk that your salary won't be at least $300,000 seems much greater than the risk of investing in a diversified stock market portfolio, which has been earning about 7 percent on the average. It is probably a lot closer to the risk of investing in penny stocks, but let's say that we would want about 15 percent for an investment of similar risk.

Now the math. We "discount" your future salary based on timing and risk to get an equivalent dollar amount in today's dollars. The formula we will use for each year n of your earnings is: incremental salary for that year divided by (1 + 15 percent risk-based return) to the nth power. So the incremental $150,000 you earn in year six is worth $150,000 / (1 + .15) × 6, or $65,000 now. The intuition behind that number is that if you invested $65,000 now in a similarly risky investment, then in six years it would be worth $150,000, the same as your salary increase. We're not done yet, however—we have to do the calculation for each of the twenty years and add them all together.

What do we get? Your future salary increase turns out to be worth $450,976 today, compared to your $500,000 cost today. If we subtract the two numbers, we get Net Present Value (NPV): in this case, -$49,000. That is the value of your investment in the MBA. Don't do it! Negative NPVs are bad investments. Not only that, but I'm about to try to convince you that NPVs as they are taught to business school students are not the right way to think about business value anyway.

Back to our product owner. She needs to prioritize user stories based on their NPV. The bigger the NPV the better. Let's try one. "As a supervisor, I would like to see how many cases are assigned to each of my account reps so I can distribute the workload better." What's the NPV? First, she can figure out the cost of the investment: the team has estimated ten story points. Of course that estimate is risky, so she'll have to account for it in the discount rate. She makes a number of assumptions and comes up with an appropriate value for that discount rate. (I'm skipping a lot of finance arguments here about whether the discount rate depends on the risk of this particular investment or the weighted average cost of capital, the return that investors demand from this company as a whole.)

Now all she needs to do is estimate the future incremental cash flows that will result from this feature. How unbalanced is the current workload? How much is that affecting revenues? How will balancing it improve revenues and costs of the company? How will it change over time? How do the competitor's actions influence it? Will the morale of the salespeople improve such that it is easier to recruit new salespeople, and therefore our hiring costs go down? In principle, all of these things can be estimated; in practice, the sprint will be long delayed as the product owner calculates the value of each story.

There is a much bigger problem with framing business value in terms of MVA and using NPV as a proxy for it: MVA is not the business value goal in the vast majority of organizations.

In the United States as of 2013, only 5,008 firms were traded on major public stock exchanges.[6] Considering the 27 million businesses in the United States, or at least the 5.7 million of them that have employees, this represents only a tiny portion.[7] For those companies, it is true, shareholder value is easily measured, and signals from the market can be used by managers to help interpret investor desires. But what about other forms of organization: privately held companies, nonprofits, and government agencies?

The remainder of the 27 million firms are private. Although many of those private companies are small, as of 2010 86.4 percent of privately held companies have more than 500 employees; these include Fidelity Investments, Cargill, Koch Industries, Toys 'R' Us, and Mars, the makers of M&Ms[8] (talk about value!). Many private companies are closely held—that is, owned by just a few owners. Family-run businesses account for 70–90 percent of global GDP. Even many large public companies are family-run (and therefore have some of the characteristics of a private, closely held company); about a third of the Global 500 companies are family-run.[9]

Some might argue that private companies should also try to maximize the market value of their owners' stakes—though it is harder to measure what that value is. Private companies can be valued, for example, through independent assessments, prices offered by potential acquirers and investors, the market value of assets held by the company, or the net present value of expected future cash flows. But the assumption that business value is tied to maximizing the financial gain of the owners does not really hold up. To see why, let me begin with an anecdote.

I was the CIO of a medium-sized private enterprise called Intrax Cultural Exchange, a company that operated cultural exchange and international education programs—high school year abroad programs, au pair programs, work and travel programs, volunteer abroad programs, and English as a foreign language schools. The company was owned and had been built without outside financing by John Wilhelm and Takeshi Yokota, who also served as the CEO. While the business as a whole was profitable, the English school line of business was a difficult one and consistently lost money. The management team reporting to Takeshi tried a number of things to improve the business, but the fundamental economics of the industry made it challenging: there were too many competitors in each local market, it was too difficult to build a brand that distinguished our schools from others, and the seasonality of the business virtually ensured that some of the schools' capacity would be unused at various times of the year. The management team—yes, including me—finally did the obvious thing: recommended to the owners that they divest that particular business line, thereby making the company as a whole more profitable.

Takeshi and John—rightly, I now must admit—were furious. They saw themselves as entrepreneurs creating new forms of international education. To them, the English schools were a critical part of the whole: without the schools, the enterprise was simply a set of disconnected lines of business permitted by US laws on cultural exchange visas. With the English schools, they had an international education business that used cultural exchange programs as a unique way to educate young people. They wanted the management team to use its creativity to make the English school business sustainable, even if it needed to lose some money. Divesting the business was not consistent with their vision.

Any notion of business value as NPV, SVA, or ROI would have had them divest the business. The moral of the story is that, as the owners of the company, they had the right to declare business value to be anything they wanted. It was their company!

When a company is publicly traded, the managers of the company cannot possibly talk to all of the owners, understand what those owners value, and then incorporate those values into investment decisions. Instead, the managers assume that SVA or MVA is a proxy for what is valued by all of the owners. But when the company is closely held, the managers *can* try to understand and apply the values of the owners. In fact, as trusted agents or fiduciaries of the owners, they *must* do so. And it turns out that—as in John and Takeshi's case—those owners do not always primarily value increases to company value or net present value of cash flows.

According to the National Venture Capital Association, in 2014 the 803 VC firms in the United States made investments in some 3,665 companies.[10] Venture capital firms have their own investment logic and their own understanding of business value. For example, an early-stage VC investor may be focused on ensuring that the company's next round of funding can be raised at a higher valuation—in other words, their biggest concern may be to ensure that the company is perceived as more valuable when it next tries to raise money, because that will cause less dilution to their ownership stake. In that case, the company creates business value by setting itself up to best match the desires of the next round's investors. And that in turn might mean maximizing market share at the expense of profits, or it might mean making investments in technologies that are trendy at the moment. It might mean recruiting a team that is trusted by investors, even at a substantial cost. Venture capital investors, based on the lifecycle stage of the fund they are managing, may also have preferences for the timeline on which their portfolio companies create an "exit" for them: either by going public or by being acquired. These considerations too may change how the company makes investment decisions.

———————

Nonprofits pose a different set of challenges. In 2014 there were 1.44 million nonprofits registered with the IRS, contributing 5.4 percent of the GDP.[11] The ultimate financial objective of a nonprofit cannot be maximizing shareholder value, since it has no shareholders. According to John Zietlow, Jo Ann Hankin, and Alan Seidner in their book on nonprofit financial management, the correct financial concern for a nonprofit should be with hitting targets for liquidity: having just enough resources to carry out the mission, but not too much.[12] But even for Zietlow, whose specialty is financial management, it would be misleading to think of business value solely in financial terms: "the public service nature of a nonprofit poses a major challenge in terms of identifying and articulating its mission and developing criteria for measuring its success."[13] The criteria for its success—that is, its definition of business value—is about accomplishing the mission for which it was chartered.

The nonprofit's mission is contained in its articles of incorporation and its bylaws, and its trustees or board of directors is legally responsible for ensuring that those documents continue to reflect the organization's mission, even if it changes. The nonprofit is expected (and required by tax authorities) to create value for both its clients and its donors by delivering on that mission. For a nonprofit formed to reduce cases of malaria in Africa, business value is not related to shareholder value or profit, but to reducing instances of malaria in Africa.

One framework for making business value decisions in nonprofits is the Dual Bottom Line Matrix.[14] The framework is a two-by-two matrix with mission impact on one axis and financial stability on the other. Projects are placed into the appropriate quadrant, which serves as the basis for making investment decisions. Projects with high impact and high sustainability are "star" projects—they clearly add value. Projects with low impact and high sustainability are valuable ("dollar signs") if they help to finance those with high impact and low sustainability ("hearts"). And those with low impact and low sustainability should be stopped or avoided.

Government agencies also have missions, and value delivery is clearly related to the performance of those missions. But many of those missions have characteristics that make it difficult to measure, project, and compare the value of investments. For example, take the Department of Homeland Security. Its mission is to keep the homeland secure. How do we measure security? If DHS is considering an investment that might—in a very small number of cases—prevent a terrorist activity, is that a value-adding activity? What if it costs $1 million? What if it costs $100 million? What if it involves a loss of privacy for citizens? What if it uses the authority of the government to compel citizens to do certain things—that is, it restricts freedom? How can we make business value decisions in such a complex, political, and emotionally charged realm?

To further complicate matters, a government agency has other value concerns beyond those directly related to its mission area. For example, the government values putting veterans to work. It values fairness in its procurement practices: all contractors and vendors should have equal chances to compete for government business. It values transparency, public accountability, and the political goals of those in power. Clearly, shareholder value is not what is meant by business value in the context of a government agency; even mission value might be rather an oversimplification.

On the other hand, there are similarities between the public sector and the private sector. As Mark Moore points out, the goal of a public sector organization can be thought of as delivering *public* value, just as that of a corporation is to deliver *private* value.[15] Both types of organizations must make value-based trade-offs in the use of resources for which there is an opportunity cost, typically cash—in one case the cash of shareholders, in the other case cash from taxpayers. The government is unique in that one of its resources is its authority to compel behavior, but doing so also has an opportunity cost. The magnitude of the punishment for non-compliance, for example, is a cost to society.[16] Moore examines several ways of thinking about public value. Is it about competently and cost-effectively delivering on the mission assigned to the agency? Possibly, but in fact agencies are generally given conflicting or incoherent guidance by their political overseers. Is it about delivering good customer service to the public? Perhaps in some cases: what kind of experience does the government deliver when you renew your driver's license? In other cases, though, customer service would be an odd way to think about government mission. Immigration and Customs Enforcement (ICE), for example, is in the business of arresting and detaining undocumented immigrants. The "customers" for their detention centers are the inmates, and increasing the comfort of their detention cell beds would be an improvement in customer service that is not necessarily a public value. Business value in a detention center is different from business value in a hotel, despite some surface similarities.

Moore concludes that all decisions about public value have to consider two broad areas: the efficient production and distribution of public goods and the fair distribution of burdens and benefits, where "fairness" depends on the decisions of politically elected representatives.[17] The decisions made by these representatives, Moore says, represent the collective aspirations of the citizenry, acting as if they were a single, collective consumer in the market.[18] Compare this to the market for private goods. The public sector generates value by creating goods and services the public is willing to pay for, just as private companies produce products that consumers are willing to pay for. The public sector also must meet political guidelines for fairness and equity in order to ensure their continued authorization by Congress, just as private companies must demonstrate their ability to create future value (future cash flows) in order to continue to receive investments from shareholders.[19]

Moore draws an interesting conclusion:

It is not enough, then, that [public] managers simply maintain the continuity of their organizations, or even that the organizations become efficient in current tasks. It is also important that the enterprise be adaptable to new purposes and that it be innovative and experimental.[20]

In other words, both private and public entities have to satisfy their equity holders that they will be able to generate future value as well as current value, and they must prove that nimbleness and responsiveness are critical to this goal and have value in themselves. Perhaps agility is already a value in the public sector!

Business managers use a variety of measures for valuing and prioritizing investments. While ROI and NPV are common choices, others include Profitability Index (Present Value of future cash inflows divided by Present Value of investment outflows), Internal Rate of Return (the discount rate that makes the NPV of the investment zero, which is compared to a minimum required hurdle rate), and Payback Period (how long until the investment recoups its costs?). Each of these measures has advantages and disadvantages, but they all share two characteristics: they are really just proxies for what the company ultimately values, like shareholder value, and they are unlikely to be useful to a product owner making feature trade-off decisions.

There is another important reason why these metrics are not so helpful in thinking about business value: they ignore the investment's role in supporting a coherent business strategy. In one of the classic works of business strategy, Michael Porter, a Harvard Business School professor and a leading authority on competitive dynamics, argues that there are really just three generic competitive strategies a business can adopt: cost leadership, differentiation, and focus. To pursue a *cost leadership* strategy, the company focuses all of its efforts on keeping its costs lower than those of its competitors. A *differentiation* strategy requires that the company provide something that is considered unique across the industry—design, brand image, feature set, technology, or dealer network, for example. With a *focus* strategy, the company zeros in on a particular target market or geographical area and serves it better than the competitors.

Importantly, Porter says that companies are unsuccessful if they try to pursue more than one of these approaches: "Effectively implementing any of these generic strategies usually requires total commitment and supporting organizational arrangements that are diluted if there is more than one primary target."[21] But if we use a simple measure like ROI to evaluate an investment, how can we make sure that it is consistent with our competitive strategy? Yes, the ROI calculation might in theory consider the effect on income that would result from undermining our generic strategy, but this is difficult to factor into the projections. More to the point, it seems backwards: we should be valuing the investment based on its contribution to strategy, not just on an income projection.

Business value, alas, is a complicated topic. Simply equating business value, customer value, and return on investment will not help, nor, as I will argue in the next chapter, will pushing the question of business value off onto a product owner. The idea that there is a single metric that represents or can serve as a proxy for business value is also misguided; in order to have a complete picture of business value, we must consider the goals of the particular organization, the interests of at least some of its stakeholders, and a variety of indicators of value, some of which may be quantifiable and some of which may not.

Avinash Dixit, in an article on options that we will be discussing later, points out that in fact, companies often make investments that they shouldn't if they are looking at NPV. "For example," he says, "entrepreneurs sometimes invest in seemingly risky projects that would be difficult to justify by a conventional NPV calculation using an appropriately risk-adjusted cost of capital."[22] Clearly there is something else going on as managers evaluate the business value of possible investments.

What is going on, I think, is that each business has a different way of understanding business value depending on its strategies, competitive situation, capabilities, mission, and people. Ultimately, if the business happens to be a public corporation, this interpretation of business value is meant to drive increases in shareholder value, but that is an empty goal until the organization translates it into specific strategies and values. And for other types of organizations—

private companies, nonprofits, government agencies—business value can be just about anything. Metrics like ROI can be interesting and useful, but they are not what we mean by business value.

If I am right about this, then business value is not a given, but rather something specific to the organization that must be discovered. This idea might sound familiar: it is in fact the very idea of agility. Agile software development starts from the assumption that it is impossible for the business to know exactly what its requirements are in advance of a software development project. Requirements must emerge; they must be discovered; an Agile process learns and adapts. I am simply saying that business value is also not a simple given at the outset of our adoption of Agile practices. Instead, business value must be discovered, must be learned, must be turned into a testable basis for valuing requirements.

Fortunately, it is there waiting to be discovered in the organization's institutional memory. Organizations have two convenient forms of institutional memory: culture and rules. In chapter 3 we will explore organizational culture and how to use it to learn what the business values; in chapter 4 we will look at rules through their most powerful instantiation: bureaucracy. The remaining chapters will turn to how we can use these learnings to create value for the enterprise.

The Meaning: the mystery deepens; we have interrogated the usual suspects but have learned little that will help us solve the case. It is curious, though, that the financial experts seem unwilling to cooperate.

1   Rappaport, *Creating Shareholder Value*. Discussion beginning on p. 59, for example, illustrates how to determine the SVA impact of alternative strategic decisions.

2   Copeland et al., *Valuation: Measuring and Managing the Value of Companies*. The authors discuss why MVA is important in "Why Value Value," pp. 22–28, and how to manage value through an example case in "Becoming a Value Manager," pp. 31–69.

3   Ibid., 3.

4   Ibid., 117.

5   Richard A. Brealey and Stewart C. Myers, *Principles of Corporate Finance*. 4th ed. (New York: McGraw-Hill, 1991), 22.

6   Dan Strumpet, "U.S. Public Companies Rise Again," *Wall Street Journal* online, February 5, 2014, http://www.wsj.com/articles/SB10001424052702304851104579363272107177430.

7   *Forbes* Online, "4 Things You Don't Know About Private Companies," reported May 26, 2013 by Mary Ellen Biery, at http://www.forbes.com/sites/sageworks/2013/05/26/4-things-you-dont-know-about-private-companies/.

8   Ibid.

9   McKinsey & Company, "Perspectives on Founder- and Family-Owned Businesses," October 2014, 4. According to a Harvard course catalog at http://www.hbs.edu/coursecatalog/1402.html, "Most companies around the world are controlled by their founders or founding families, including not only private firms but also more than half of all public corporations in the U.S. and Europe, and more than two thirds of public corporations in Asia."

10 National Venture Capital Association Yearbook 2015 (Thomson-Reuters).

11 McKeever, Brice S. and Sarah L. Pettijohn, "The Nonprofit Sector in Brief 2014" Urban Institute. October 2014. Online at http://www.urban.org/sites/default/files/alfresco/publication-pdfs/413277-The-Nonprofit-Sector-in-Brief--.PDF.

12 John Zietlow, Jo Ann Hankin, and Alan Seidner, *Financial Management for Nonprofit Organizations: Policies and Practice* (Hoboken: Wiley, 2007), 195.

13 Ibid., 6.

14 Discussed in Zietlow et al., 327, who credit it to Jeanne Bell Peters and Elizabeth Schaffer in Financial Leadership for Nonprofit Executives. I have also found this matrix appearing without credits, for example in David Renz and Robert D. Herman, *The Jossey-Bass Handbook of Nonprofit Leadership and Management* (John Wiley and Sons, 2010), 466. There may be some confusion: there is a different Dual Bottom Line Matrix commonly referred to in literature about socially responsible businesses.

15 Mark H. Moore, *Creating Public Value: Strategic Management in Government* (Cambridge, MA: Harvard University Press, 1995), 28.

16 Ibid., 42.

17 Ibid., 48.

18 Ibid., 30.

19 Ibid., 53.

20 Ibid., 55.

21 Michael Porter, *Competitive Strategy: Techniques for Analyzing Industries and Competitors* (New York: Free Press, 1985), 35.

22 Avinash K. Dixit and Robert S. Pindyck, "The Options Approach to Capital Investment" in *Harvard Business Review*, May–June 1995, p. 109.

How did I get into the world? Why was I not asked about it and why was I not informed of the rules and regulations but just thrust into the ranks as if I had been bought by a peddling shanghaier of human beings? How did I get involved in this big enterprise called actuality? Why should I be involved? Isn't it a matter of choice? And if I am compelled to be involved, where is the manager—I have something to say about this. Is there no manager? To whom shall I make my complaint?

Søren Kierkegaard, *Repetition*

If this world is a poem, it is not because we see the meaning of it at first but on the strength of its chance occurrences and paradoxes.

Maurice Merleau-Ponty, *Signs*

3

THE CULTURE

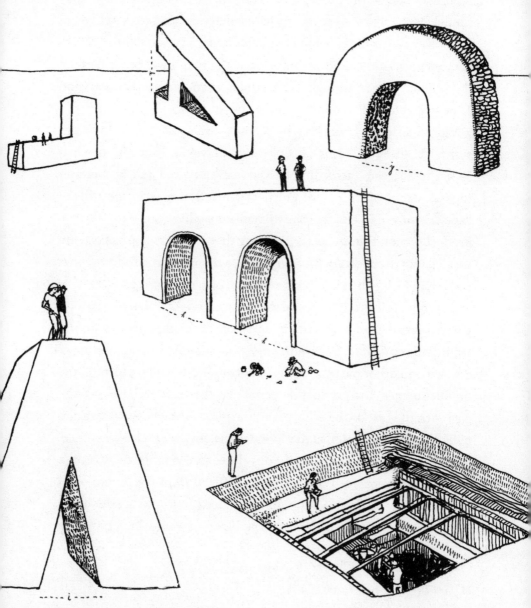

$2h + i + j + k = 52 \ Armadillos$

I have said that business value is not a given, but rather that it must be discovered. Is this really the business of the Agile team? Or is it the business of the business, and the business of the Agile team to deliver efficiently on it once the business decides what it values? Isn't it the role of the product owner[1] to understand business value and apply it in creating and prioritizing user stories?

Product owner, of course, is a Scrum-specific role. But other Agile frameworks rely to a greater or lesser extent on representatives of the business to indicate what is valuable and what is not. Since Scrum is most prescriptive and definite about the role, I will focus here on the product owner, though I think there is an analogous discussion to be had about XP and other frameworks.

And prescriptive Scrum is. The product owner must be a single person with authority over all questions of value. "For the product owner to succeed," according to Ken Schwaber and Jeff Sutherland, the founders of Scrum, "the entire organization must respect his or her decisions."[2] To ensure that this responsibility rests with a single product owner, they go as far as to say that "the Development Team *isn't allowed* to act on what anyone else says."[3] Scrum's product owner is the OPYCLT—the *Only Person You Can Listen To*.

This model of product ownership is curiously similar to a design pattern we are familiar with in software development. The Scrum team is *encapsulated*; it is self-contained, self-organizing, and interacts with all other parts of the system through a single interface: the product owner. The product owner allows the team to be *loosely coupled* with the rest of the enterprise. The main advantage of loose coupling in software design is that it hides the details of *implementation*: as long as the product owner (or interface) can be relied on, the organization does not need to know the details of what the Scrum team is doing, and the Scrum team does not need to know the details of how the organization is determining business value. This gives the

Scrum team freedom to self-organize: the product owner frames the business problem, and the Scrum team provides the solution (that is, the implementation). Perhaps the analog here is really a firewall: once the business tosses the business problem over the firewall—so to speak—to the Scrum team at the beginning of the sprint, a Scrum Master "protects" the team so it can find its own solution to the issue.

It is an intuitively appealing model to those of us in the software business, but I believe that it is fundamentally flawed.

In *X-Teams: How to Build Teams that Lead, Innovate, and Succeed*, Debora Ancona and Henrik Bresman propound the idea of an X-team, a team with both an internal and an external focus, as the best model for dealing with the complexities of a business organization in a competitive market. "While managing internally is necessary," they say, "it is managing externally that enables teams to lead, innovate, and succeed in a rapidly changing environment."[4] An X-team does not just passively receive business needs and constraints "tossed over the wall" from the rest of the business; on the contrary, "X-teams seek out information about the customer (often directly as opposed to secondhand), the technology, the market, and the competition."[5] Ancona and Bresman's research indicates that the success of a team requires almost the *opposite* of encapsulation:

High performing teams manage across their boundaries, reaching out to find the information they need, understand the context in which they work, manage the politics and power struggles that surround any team initiative, get support for their ideas, and coordinate with the myriad other groups that are key to a team's success.[6]

By pushing beyond their boundaries and interpreting the business's needs (and values) themselves, X-teams are able to achieve higher levels of performance:

Teams that scouted out new ideas from outside their boundaries, received feedback and coordinated with outsiders, and got support from top managers were able to build more innovative products faster than those that dedicated themselves solely to efficiency and working well together.[7]

In the X-team model, information flows in two directions through multiple interfaces. Not only does the team absorb information about the business, its competitive environments, its politics, and its strategies, but it also acts as an ambassador of its own ideas and work.

Once we abandon the idea that business value is a simple formula, known ahead of time and simply applied to user stories, and accept instead that business value requires interpretation, we have to ask whether the model of a product owner doing all the interpretation continues to make sense. An X-team, a team that has multiple touch points with the organization and even influences the organization's understanding of what it *should* value, might be more appropriate. In fact, it's not clear that the encapsulation model is ever really used in practice. At the very least, the IT organization is providing a set of requirements in the form of technical best practices and compliance requirements that the product owner does not "own."

Many Scrum teams struggle with getting the product owner to prioritize non-functional requirements, such as security stories and work that will reduce technical debt, alongside functional user-facing requirements. One often hears the argument that the product owner *should* be able to prioritize those requirements—after all, they are a matter of business value—so it is merely a matter of communicating their business value to the product owner. In my experience, security requirements and other non-functional requirements tend to make their way into sprints as the result of a negotiation between

IT management and the product owner or as the result of under-the-radar work that the team does just as a part of delivering the functional requirements because they consider it a quality attribute of the functional code.

Is the product owner model actually an *effective* way to manage those non-functional requirements? The product owner is given control over business value trade-offs because he is an expert on business needs; is it reasonable to expect that his expertise extends to security needs, code maintainability, and performance engineering? Sure, the team might be able to explain a security need, but does the product owner have the intuitive feel for it to prioritize it against functional requirements? I believe that the product owner model is based on a hidden assumption: that both the functional requirement and the security requirement can be translated into ROI terms and compared apples-to-apples. If business value is a much more subtle concept than ROI or any other financial metric, then we are asking too much of the product owner.

In fact, security is not *just* a set of requirements or user stories that a product owner has on a backlog and prioritizes along with functional requirements. It is about risks and mitigations, about a strategic approach to risk management across the business. Risk management is an aspect of how the enterprise interprets business value, and it becomes part of a set of strategies created by security experts to maximize that value. The natural owners of that strategy—the ones who can make the best value decisions concerning risk and its mitigation—are generally from an entirely different part of the business than the product owner. The result of this, as we will see in the next chapter, is rule-making and bureaucracy, as the security team tries to impose this strategy on product owners. X-teams might be a more organic way to bring the owners of the risk management strategy into the process of value creation. It's true that the product owner can incorporate the input of the security group, but that reduces the interactivity that Agile teams value.

It is not just Bresman and Ancona who believe that teams should be proactive interpreters of their environments. Richard Hackman, for example, argues that "[team] members should take an active, rather than a reactive stance toward the environment in which the team operates, continuously scanning the environment and inventing or adjusting their performance strategies accordingly."[8] As with Ancona and Bresman, he is not talking specifically about software development teams. But why would software development teams be different? I view it as the consequence of an outdated belief that IT organizations are the passive recipient of business requirements from *the business*, and judged not on outcomes but on their success in delivering what they are told is needed. If we really want the Agile team to be responsible for delivery of business value, we need to give the team ownership over business value discovery and interpretation, not just over delivery on requirements.

In my mind, Eric Ries's *Lean Startup*[9] is a landmark not only for its view of how to launch new products through validated learning but also for the view of business value that underlies it. In my reading, Ries is suggesting that the business does not even know what will generate business value; instead, it entertains a series of hypotheses about value and then attempts to confirm or disprove them, arriving at a finer and finer understanding throughout the process. Think of how far this is from the idea that we prioritize features by projecting future cash flows and returns, applying a formula, and comparing the "objective" measures that result. For Ries, the cash flow projections are at best a hypothesis, and his approach involves finding the most important drivers of those cash flows and designing experiments that will produce validated learning about them.

Ries's approach is rightly focused on *user* or *customer* value, since that is a large area of risk and uncertainty in the formulation of business value for a startup. But we need to expand his ideas to include other notions of business value. If business value might imply deliberately losing the least profitable 10 percent of customers, the team might begin with a model of the business with those 10 percent of customers gone and validate it with the CFO's office. They might then experiment with different feature changes to determine which preserve the desired customers and drive away the unprofitable customers. Based on customer behavior, they can see whether losing the unprofitable customers increases value for the company, or they can pivot and try another approach to value generation.

In this formulation, the product owner is no longer the interpreter of business value. Instead, value is discovered by the team that does the validated learning experiments and hypothesis testing. There is no need for a single interpreter, since the entire team can learn together. The findings of the team then influence how the remainder of the organization thinks about business value. This is the model that DevOps points toward: an inclusive model where the team has all the skills necessary to create value.

In the traditional model where the Agile team, with the guidance of its product owner, knows how to achieve business value by optimizing a simple metric, the rest of the organization should step aside and let the team do its work. If, on the other hand, the meaning of business value is something that must be unearthed by the team, then the organization as it exists now may contain vital clues. Unfortunately, the Agile literature does not always share that interpretation.

"Adopting Scrum," say Larman and Vodde, "requires change in the *whole* organization."[10] And not just superficial change: they also cite another management theorist, Daniel Meyer, arguing that "successful implementation of multifunctional teams requires a *fundamental redesign of the entire organization*."[11] There is a strong fist-pounding, "this must change" strain in the Agile literature that thinks of the organization as it currently exists as an impediment or obstacle to Agile practice. Jeff Sutherland, in his own retrospective of Agile practice ten years after the Manifesto, says that "organizations must be structured for Agile response. Failure to remove impediments that block progress destroys existing high-performing teams and prevents the formation of new high-performing teams."[12] Perhaps typical of the discourse is Mario Moreira's claim in *Being Agile*:

When we discuss Agile adoption, we are talking about a change to the organizational culture. Culture disruption is never painless. This is because adopting Agile is not a matter of learning skills or understanding a procedure; it is about adopting a set of values and principles that require change in people's behavior and the culture of an organization.[13]

This way of thinking has always struck me as a little strange. Our goal is to deliver value, to figure out how to meet the needs that are determined by the organization, and yet we consider the organization to be the biggest impediment to doing so. The only explanation I can think of for this is that we are implicitly assuming that there is a stable, objective, preordained definition of business value, and we are determined to deliver on that definition *despite* the organization around us. In my experience, this arrogance is not warranted; in fact, the organization probably understands value in ways that the Agile team does not, and the obstacles to Agile adoption actually tell us something useful about business value in the organization.

There is a better way. Instead of sweeping away corporate culture and existing organizational design in a "big bang," we can move incrementally, learning as we go. We can take an Agile approach to Agile adoption, driving organizational change incrementally based on what we can learn about the organization's true needs and dynamics. Ken Schwaber talks about using an Enterprise Transition Backlog in *The Enterprise and Scrum*. Christopher Avery, in his work on organizational change, talks about adopting an Agile approach to Agile adoption. Citing the early twentieth-century organizational psychologist Kurt Lewin, he argues that we can never understand an organization until we try to change it and thus suggests a "provoke and observe" approach:

We can never direct a living system, only disturb it and wait to see the response . . . We can't know all the forces shaping an organization we wish to change, so all we can do is provoke the system in some way by experimenting with a force we think might have some impact, then watch to see what happens.[14]

Avery's "provoke and observe" approach parallels the Agile principle of "inspect and adapt." In my interpretation of Avery's comments, we can formulate a hypothesis about what is valuable to the organization, deliver something based on that hypothesis, and observe the results. The feedback that results—especially if this is a new way of thinking about value—teaches us how to either adjust our hypothesis or adjust our way of bringing agility into the culture. In terms of the Lean Startup, you may see this as an example of adjusting our value hypothesis or our growth hypothesis through validated learning.

The organization's culture has evolved for a reason: it has been strongly influenced by what the organization has found to deliver business value in the past. Edgar Schein has written extensively on corporate culture. He explains culture as the organization's interpretation of what behaviors are effective:

Most organizations evolve assumptions about their basic mission and identity, about their strategic intent, financial policies, fundamental way of organizing themselves and their work, way of measuring themselves, and means for correcting themselves when they are perceived to be off target.[15]

These organizational assumptions—which in my interpretation express the organization's understanding of business value—become the corporate culture.

Culture is a pattern of shared tacit assumptions that was learned by a group as it solved its problems of external adaptation and internal integration that has worked well enough to be considered valid and, therefore, to be taught to new members as the correct way to perceive, think, and feel in relation to those problems.[16]

The organization has had to learn what business strategies, values, protocols, and behaviors work in its environment to support its ultimate aims, whether those are maximizing shareholder value or accomplishing mission objectives. That learning forms the basis of tacit assumptions and norms, the organization's collected wisdom about what behaviors foster success. And if success means accomplishing the ultimate goals that serve as the sources of business value, then the Agile team must come to understand those values, strategies, goals, and operational modes that are embedded in the culture around it—that is, the business values that have been known to foster success. What are those elements of business value?

For most business organizations, financial performance is the primary error-detecting mechanism because of its seeming objectivity, but cultural assumptions dominate even what kind of information is gathered and how it is interpreted. For example, some companies go almost exclusively by the stock price as the indicator of how they are doing. Others look at debt-to-equity ratios, cash flow, or market share. In each case, cultural assumptions arise from the indicators that work best. If the organization is functionally organized, it may also develop a subculture around the finance function, and actual conflicts may develop between finance, production, engineering, and marketing over which indicators to use in assessing company performance.[17]

The difficulty is that culture is unwritten and unstated. For the Agile team to use culture as a guide to what is valued by the business, it must excavate, deduce, hypothesize, and test. One thing that it cannot do is to try to rip out the existing culture and replace it with a "more Agile" one.

Corporate culture is not an impediment to Agile adoption; it is a valuable clue to defining the success of Agile adoption.

In a way, this is a question of respect and inclusion. Rather than assuming that we have a better way and the organization will just have to change to support it, we can assume that the organization is bringing something to the table in its culture, strategies, and processes. The organization has unique goals, a unique competitive environment, and unique resources and competitive advantages. Based on those things, it has developed a culture and a set of processes that represent an interpretation of what will create business value. What we must do in adopting an Agile practice is to both change and absorb that culture and those processes through an inspect-and-adapt approach.

Agile approaches attempt to bring together developers and *the business* in an atmosphere of mutual respect and joint contribution. Until now, however, the focus has been on users of the software, product visionaries, and developers. Recent developments in the Agile world—notably DevOps—have broadened this idea of respect and inclusion to encompass Operations and Security. The DevOps model, in other words, looks to break down the silos that have resulted from technical specialization over the last few decades. But the DevOps spirit goes further, looking to eliminate the conflicting incentives of organizational silos and the inhumane behaviors that can result from those conflicting incentives.

Perhaps we can take this idea even further still. There is no reason why the DevOps team's responsibility needs to stop at the border of what used to be considered IT. The team is part of a broader enterprise, whose collective knowledge, skills, and judgment need to be part of the value creation process.

Encapsulation of the software delivery team is a poor fit with DevOps and contemporary Agile culture.

The Culture: another shadowy player emerges, with the mysterious name "X-team." Following a lead from our informant Schein, we suspect that the organization is hiding something. Sure enough, we are able to piece together a rough sketch of the culprit by asking the right questions.

1 Again, I am using the Scrum-focused term "product owner," but I mean to include any decision process for making business value decisions and trade-offs.

2 Ken Schwaber and Jeff Sutherland, "The Scrum Guide: The Definitive Guide to Scrum: The Rules of the Game," PDF at scrum.org., October 2011, p. 5.

3 Ibid., 6. Emphasis is mine.

4 Debora Ancona and Henrik Bresman, *X-Teams: How to Build Teams that Lead, Innovate, and Succeed* (Boston: Harvard Business School Press, 2007), Kindle loc. 105. Larman and Vodde, in *Scaling Lean and Agile Development*, pp. 202–203, also refer to Ancona and Bresman (and Hackman—see below), mostly in the context of removing project managers from the equation and giving the team access to the rest of the organization. I completely agree, but I also believe this idea goes much deeper.

5 Ibid., loc. 110.

6 Ibid., loc. 667.

7 Ibid., loc. 267.

8 Richard J. Hackman, *Leading Teams: Setting the Stage for Great Performances* (Boston: Harvard Business School Press, 2002), 106.

9 Eric Ries, *The Lean Startup: How Today's Entrepreneurs Use Continuous Innovation to Create Radically Successful Businesses* (New York: Crown Business, 2011).

10 Craig Larman and Bas Vodde, *Scaling Lean and Agile Development*, 233. Emphasis is mine.

11 Ibid., 208. Emphasis is mine.

12 Jeff Sutherland, "Ten Year Agile Retrospective: How We Can Improve in the Next Ten Years." Microsoft Developer Network (MSDN), https://msdn.microsoft.com/en-us/library/hh350860(v=vs.100).aspx. Larman and Vodde quote Sutherland expressing similar sentiments on p. 230.

13 Mario Moreira, *Being Agile: Your Roadmap to Successful Adoption of Agile* (New York: Apress, 2013), 8.

14 Christopher Avery, "Responsible Change." Cutter Consortium Agile Project Management Executive Report 6 (10): 1–28. 2005. pp. 22–23.

15 Edgar H. Schein, *The Corporate Culture Survival Guide*. Revised ed. (San Francisco: Jossey-Bass, 2009), 40.

16 Ibid., 27.

17 Ibid., 49.

One must lie low, no matter how much it went against the grain, and try
to understand that this great organization remained, so to speak,
in a state of delicate balance, and that if someone took it upon himself
to alter the dispositions of things around him, he ran the risk of losing
his footing and falling to destruction, while the organization would
simply right itself by some compensating reaction in another part of
its machinery—since everything interlocked—and remain unchanged,
unless, indeed, which was very probable, it became still more rigid,
more vigilant, severer, and more ruthless.

Franz Kafka, *The Trial*

Each problem that I solved became a rule, which served afterwards
to solve other problems.

René Descartes, *Discourse On Method*

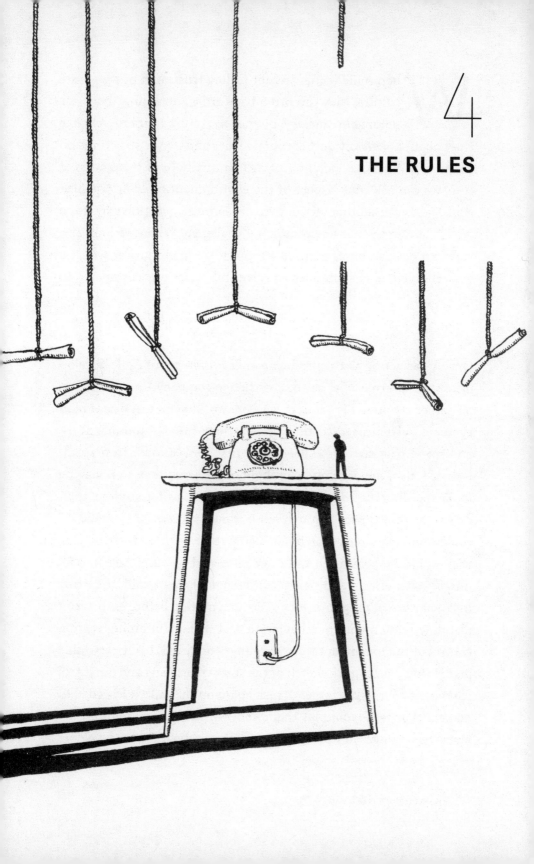

4

THE RULES

W hen Agile teams are not feeling frustrated by corporate culture, they can often be heard complaining about corporate bureaucracy. But just as culture contains valuable clues about business value, corporate rules and power structures do so as well. When we look at a particular attribute of corporate culture, we can ask what aspect of the environment or what organizational goals caused that attribute to develop and question why it is a good solution to that environmental challenge. When we look at a rule or a piece of bureaucratic machinery, we might just as well ask what problem it was intended to solve and why it was considered to be a good solution.

Once I was acting as the product owner for a team of Agile developers at the international cultural exchange company I mentioned in an earlier chapter. The product we were working on was a case management system used to track applications for our summer work-and-travel program for college students from abroad. In a sprint retrospective, the developers brought up an issue they were having understanding the complicated status transitions for student applications as they flowed through our business process. They asked if I could provide more complete documentation, making the case that they would be able to write better tests and be sure they weren't missing side effects and pieces of the logic. I was afraid of getting into a pattern of preparing lots of documentation before each sprint planning session and then having to deal with maintaining version control of that documentation as changes occurred. I suggested this plan: I would do a quick sketch of the state transitions and pin it to a corkboard next to the project's task board on the wall. If I needed to make a change, I would let them know in the daily standup, then make the change and pin it back to the corkboard. If a developer needed the information, she could pull it from the corkboard and replace it when she finished.

The team was satisfied, but a problem emerged in the next sprint retrospective. During the sprint, one of the developers had taken the sketch off the corkboard to use and then failed to return it. That caused some confusion when other developers weren't able to find it. One of the developers proposed a solution: there would be another piece of paper on the corkboard that was a sign-in/sign-out sheet for the state transition sketch. When a developer took down the sketch, he would write his name on the sheet, along with the date and time, and then enter the date and time he returned it. Since we might wind up with multiple requirements documents pinned to the board, the developer would also note which document he was signing out. That would fix the problem of last sprint.

I offered an alternative suggestion: how about if each developer would just remember to return the sketch to the corkboard in a reasonable time?

If this story makes sense to you, then you understand both continuous improvement and bureaucracy. The developer was solving a legitimate problem by creating a new process, and the process required making a rule—a rule that would impose a cost on the developers. This is one of the most salient characteristics of bureaucracy. The rule happened to involve producing a document, which is also quite common in bureaucracy. It was a rule that restricted individual judgment. You might say that it transferred responsibility, since as long as the developers obeyed the rule—signing out the sketch—it was less important whether they were considerate and returned the sketch promptly.

It sounds strange to say this, but developers are bureaucrats by nature. We have a tendency to solve problems by creating standard processes rather than by relying on human judgment. "If you break the build, you must revert your code and fix the problem—immediately!" That is the voice of the bureaucrat. We can hear that same bureaucratic ring in a DevOps rule that says, "Developers must manage their own code in production for six months, after which an operational readiness review is held and the operations team decides whether to accept responsibility for it." Recent trends in software development have moved us even closer to bureaucratopia: we now consider it part of the developer's job to write tests that "prove" that his or her code does what it is supposed to do. Doesn't that sound like the mounds of paper devoted to ensuring that the bureaucrats have followed the appropriate processes—the required signatures and routing sheets and checklists? Bureaucracy is all about rules and proving that they have been followed. I know that if you are developer, then a bureaucrat is exactly what you don't consider yourself—but what I am really trying to say is that there is nothing wrong with bureaucracy *per se*.

Large organizations tend to steer toward bureaucracy as a way to maintain standards of behavior. Hating bureaucracy is a bit like hating the law of entropy. It is simply a fact, a method of imposing order on chaos (okay, that makes it the opposite of entropy). That doesn't mean that there isn't much to hate about poor bureaucracy or poor uses of bureaucracy. The point that I will make here is that bureaucratic behavior generally follows from a business need, and the only way to remove or improve the bureaucracy is to find a better way to meet the underlying need. In this peculiar sense, bureaucracy can be said to add business value—it meets a business need (though note that it is not a *user* need or a *customer* need). Of course, a particular bureaucratic process might not add as much business value as a different solution might. The purpose of an Agile team is to self-organize and meet the *underlying* business need in the best way possible, often by cutting through the bureaucracy.

———————————

To learn more about bureaucracy, we can do no better than to look at the federal government. At this point I'd like to introduce Max Weber, the author of *Economics and Society* (first published in 1922, shortly after his death), whom we will be discussing extensively in this section. Weber is often credited with being one of the founders of the field of sociology; his goal was to describe the logic behind different social structures, which he often found to be based on economic factors. Although Weber is often cited as someone who—perhaps surprisingly—believed that bureaucracy was a good thing, what he was really driving at was that bureaucracy is a *rational* solution to meeting certain societal needs.

While Weber had a lot to say about government bureaucracy, he also makes it clear that "this type of organization is in principle applicable with equal facility to a wide variety of different fields. It may be applied to profit-making business or in charitable organizations, or in any number of other types of private enterprises serving ideal or material ends."[1] Bureaucracy is everywhere—arguably even in sprint retrospectives, though Weber was not familiar with them. Schein also points out that a normal part of organizational growth is the creation of rules and paperwork to make up for the loss of face-to-face contact.

Weber's positive take on bureaucracy is startling to those of us for whom the word "bureaucracy" has such negative connotations. In *Economics and Society* we read:

Experience tends universally to show that the purely bureaucratic type of organization—that is, the monocratic variety of bureaucracy—is, from the purely technical point of view, capable of attaining the highest degree of efficiency and is in this sense formally the most rational known means of exercising authority over human beings. It is superior to any other form in precision, in stability, in the stringency of its discipline, and in its reliability. It thus makes possible a particularly high degree of calculability of results for the heads of the organization and for those acting in relation to it. It is finally superior both in intensive efficiency and in the scope of its operations and is formally capable of application to all kinds of administrative tasks ... the choice is only between bureaucracy and dilettantism in the field of administration.[2]

It's not just Weber with this peculiar view. Sociologists Daniel Katz and Robert Kahn claim that "[bureaucracy] is an instrument of great effectiveness; it offers great economies over unorganized effort; it achieves great unity and compliance."[3] "Indeed," say Daniel Wren and Arthur Bedeian in *The Evolution of Management Thought*, "almost all the benefits we take for granted in today's society—modern medicine, modern science, modern industry—rest on a bureaucratic foundation."[4]

How could anyone possibly make claims like these for the soulless system that we know as bureaucracy?

It helps to understand what bureaucracy was meant to replace. When bureaucracy entered politics, it was seen as an improvement on arbitrariness, capriciousness, nepotism, and other undesirable characteristics of public administration. Bureaucracy would ensure fairness by applying rules to administrative behavior. The rules would be the same for all cases—no one would receive preferential or discriminatory treatment. Not only that, but the rules would represent the best products of the accumulated knowledge of the organization: formulated by bureaucrats who were experts in their fields, the rules would impose efficient structures and processes while guaranteeing fairness and eliminating arbitrariness.

In Wren and Bedeian's explanation, bureaucracy is what results when we replace personal power and capricious activity with administration through *legal authority*.[5] They list the advantages of bureaucracy:[6]

Characteristic	Advantage
Division of labor	Efficiency through specialization
Managerial hierarchy	Clear chain of command
Formal selection	Based on merit and expertise
Career orientation	Freedom from external pressures
Formal controls	Efficiency
Impersonality	Protection from arbitrariness

The essence of bureaucracy is that its rules are applied impartially—or even more to the point, impersonally. As Weber puts it, the rules should be applied *sine ira et studio*—that is, "without anger or bias."[7] The conclusion he draws from this is also surprising. "'Objective' discharge of business primarily means a discharge of business according to calculable rules and 'without regard for persons' . . . 'without regard for persons' is also a watchword of the market and all naked economic interests."[8] In other words, bureaucracy is consistent with—one might even say demanded by—the capitalist economy. As with bureaucracy, the invisible hand of the market, a core principle of capitalism, sweeps all human concerns aside and is a purely mechanical force. Firms can do business better knowing that the bureaucracy will treat them according to rules: they can predict what the results of their actions will be.

Bureaucracy develops . . . the more it is "dehumanized," the more completely it succeeds in eliminating from official business love, hatred, and all purely personal, irrational, and emotional elements which escape calculation. This is appraised as its special virtue by capitalism.[9]

Somehow Weber has taken us from bureaucracy as an obstacle to bureaucracy as a social force for good!

Of course, Weber and other authors are aware of the problems with bureaucracies. Rules become set in stone and can't change with circumstances. Rigidity discourages innovation. Rules themselves come to seem arbitrary and capricious. Their original purpose gets lost and the rules become goals rather than instruments. Bureaucracies can become demoralizing for their employees.

My point is not about whether bureaucracy is good or bad. It is rather about the real business needs that are served through bureaucratic behavior. A government needs to ensure fairness and eliminate arbitrariness. The solution it has chosen is bureaucracy. If we don't think bureaucracy is the best solution, we are still left with the naked business need. The developers need to make sure the state transition sketch is on the corkboard when they need it. A bureaucratic rule is one possible solution. There may be better solutions.

When you "provoke and observe" in the government environment, you discover what its true values are. The US government is based on a system of "checks and balances"—in other words, a system of distrust. The great freedom enjoyed by the press, especially in reporting on the actions of the government, is another indication of the public's lack of trust in the government. As a result, you find that the government places a high value on transparency. While companies can keep secrets, government is accountable to the public and must disclose its actions and decisions. There is a business need for continued demonstrations of trustworthiness, or we might as well say a business value assigned to demonstrating trustworthiness. You find that the government is always in the public eye—the press is always reporting on government actions, and the public is quick to outrage. Government agencies, therefore, place a business value on "optics"— how something appears to the observant public. In an oversight environment that is quick to assign blame, government is highly risk averse (i.e., it places high business value on things that mitigate risk). Provoking and observing teaches us what the government values, and it gives us a more complete view of the business needs that will yield business value when delivered. Delivering business value in the government means meeting these business needs.

When an Agile team self-organizes to meet business needs and deliver business value, it cannot just consider customer and user needs for its products. It must consider all of the needs of the organization and all of the things that the business values, and then self-organize to meet all of those needs. The needs disclosed by bureaucratic rules are among those needs.

Bureaucracy delivers business value. Just sometimes not enough.

Bureaucracy in an organization often leads to *compliance* requirements, which in turn generate work items that must be prioritized alongside functional requirements. It is easy to mistake these requirements for *waste*. Waste is whatever does not add value, and compliance requirements are not concerned with adding direct customer value. This explains the frustration that teams often have when confronted with bureaucracy.

But what if those requirements actually are adding business value— just an indirect and well-disguised type of value? If the public requires transparency into a government IT project, then activities to provide that transparency are not necessarily waste. If the government values fair competition between vendors and supporting veterans through preferential hiring practices, the additional process steps those concerns add to our IT delivery value chain are not necessarily waste, though they do not add user value. If the public demands independent validation of the results of a government IT project, and as a result there is independent duplication of testing that has already been performed, then that is not necessarily waste. It is simply a type of shareholder value, a demand that the "owners"—or at least key stakeholders of the government—demand as one of the returns on their investment of tax money.

———————

Compliance requirements can be found in most organizations. Some of these requirements, like those of the Sarbanes-Oxley Act and the Health Insurance Portability Act (HIPAA), come by way of the government. Others, like the requirements of the Payment Card Industry (PCI), are imposed by non-governmental organizations. Some are imposed by auditors. Some are self-imposed. In theory, these requirements can be translated into Shareholder Value Added, Net Present Value, or Return on Investment because they affect the amount or risk of future cash flows. Meeting PCI requirements allows companies to accept credit cards, which probably increases revenues. Getting a clean audit builds trust in the company and affects its ability to raise capital, thereby supporting its market value. So we can be pretty sure that there are questions of business value involved in compliance.

Security requirements are often expressed in terms of compliance. Perhaps this is because they come from a different part of the organization, the part concerned with overall risk management strategy, which does not have direct control of risk on a project-by-project basis. As a result, the security "bureaucracy" issues rules that represent their best approach to generating business value through risk control, and projects are expected to comply. Security concerns may be especially difficult to translate into simple business metrics at a granular level: what, for example, is the business value of opening or closing a particular firewall port? A DevOps approach that makes security a direct concern of the team can provide a better solution than this arms-length imposition of security requirements. But we must be careful to note two things. First, the compliance requirements are not an obstacle, but rather an expression of a deeper business need that the team must still address. Second, the team does not necessarily go its own way with security strategy; it works within parameters set by the security leadership, which interprets the ultimate goals of the organization into specific risk-management strategies. We will dive further into the interpretive role of management in the last chapter.

Let's take a final example from a very different part of the organization. The Marketing function often has branding guidelines as well as style guidelines to support the brand. These may be introduced as compliance requirements, not just for look and feel of software but also the flow of user interactions. A brand, according to Philip Kotler, "is essentially a seller's promise to consistently deliver a specific set of features, benefits, and services to the buyers. The best brands convey a warranty of quality."[10] Brands have *equity*; that is, they are a component of shareholder value. A brand is worth money because it reduces marketing costs for products since customers are already familiar with the company, because the higher perceived quality lets the company charge more for its products, and because it gives the company negotiating leverage with distributors who are expected to carry the brand. Brands help create customer loyalty that in turn drives future cash flows.

There is a cost to managing a brand as well. The company needs consistency in its packaging and its communications, for example. It needs consistency in the benefits it offers to customers and in its style of interactions with customers. For the Agile team, this may translate into compliance requirements. It should be clear by now that these compliance requirements, far from being obstacles, are actually a source of business value.

Strategies and best practices can be embodied in rules. When and how do those rules become bureaucracy?

In one sense, they already are. According to Weber, "Bureaucratic administration means fundamentally the exercise of control on the basis of knowledge."[11] Similarly, John Stuart Mill considered bureaucracy just a form of administration that "accumulates experience, acquires well-tried and well-considered traditional maxims, and makes provision for appropriate practical knowledge in those who have the actual conduct of affairs."[12] In other words, it is simply a form of administration that captures knowledge in rules and makes it available to experts to apply. In a thought-provoking article for *Harper's Magazine*, David Graeber argues that what we think of as government bureaucracy is actually just an inheritance from corporate rule making:

Americans often seem embarrassed by the fact that, on the whole, we're really quite good at bureaucracy. It doesn't fit our American self-image . . . if Americans are able to overlook their awkward preeminence in this field, it is probably because most of our bureaucratic habits and sensibilities—the clothing, the language, the design of forms and offices—emerged from the private sector.[13]

"Consider," he says, "the maze of rules one must navigate if something goes even slightly awry with a bank account." Interestingly, Karl Marx also expressed displeasure with bureaucracy—also because he viewed it as a quality of corporations and therefore capitalism. "The Corporation is civil society's attempt to become state; but the bureaucracy is the state which has really made itself into civil society."[14]

While disavowing rules, the Agile community is actually full of them. This is understandable, because rules are a way of bringing what is considered best practices into everyday processes. What would happen if we made exceptions to our rules—for instance, if we entertained the request: "John wants to head out for a beer now, instead of fixing the problem that he just introduced into the build?" If we applied the rules capriciously or based on our feelings, they would lose some of their effectiveness, right? That is precisely what we mean by *sine ira et studio* in bureaucracy. Mike Cohn, for example, tells us that "improving technical practices is not optional."[15] The phrase *not optional* sounds like another way of saying that the rule is to be applied "without anger or bias." Mary Poppendieck, coauthor of the canonical works on Lean software development, uses curiously similar language in her introduction to Greg Smith and Ahmed Sidky's book on adopting Agile practices: "The technical practices that Agile brings to the table—short iterations, test-first development, continuous integration—are not optional."[16] I've already mentioned Schwaber and Sutherland's dictum that "the Development Team *isn't allowed* to act on what anyone else [other than the product owner] says."[17] Please don't hate me for this, Mike, Mary, Ken, and Jeff, but that is the voice of the command-and-control bureaucrat. "Not optional," "not allowed,"—I don't know about you, but these phrases make me think of No Parking and Curb Your Dog signs.

I understand where these kinds of statements come from. These pioneers and leaders of the Agile world have seen people make mistakes in exactly these areas, and they (rightly) know better. The rules are a form of institutional memory: the Agile community has learned what works, and these rules represent that learning. In fact, as memory goes, rules are a great improvement over documents and hard drives—the rules are actually executable. Just as tests are executable requirements, imperatives from the Agile leaders are executable best practices.

Even in the document that set off the train of thought that eventually became the Scrum methodology, "The New New Product Development Game," Hirotaka Takeuchi and Ikujiro Nonaka point out that "knowledge is also transmitted in the organization by converting project activities to standard practice . . . Naturally companies try to institutionalize the lessons derived from their successes." However, they warn that "institutionalization, when carried too far, can create its own danger." That danger, of course, is petrification, the result we often see with bureaucratic rules.

Are these best practice rules really bureaucracy? The problem with bureaucracies is that the rules sometimes are not allowed to change when the best practices change. That petrification of rules is what frustrates us in the bureaucracy. But in his study of Toyota-influenced Lean manufacturing at the NUMMI auto plant, Paul Adler distinguishes between "compliance bureaucracies" and "learning bureaucracies:"[18] the latter allows rules to evolve based on worker input. How quickly, I wonder, will Schwaber and Sutherland's rule change if we as an Agile community no longer think that the Scrum team must only listen to the product owner? What if test-first development turns out not to be a great idea, or if another practice inconsistent with it turns out to be even better? Poppendieck said that it was not optional!

Perhaps Poppendieck doesn't really mean that the rule is fixed. She has created the rule because there is an underlying need or a set of underlying needs, and test-first development satisfies those needs. Schwaber and Sutherland formulated their rule because they are concerned with an underlying need (an empowered person who can make value trade-offs). Sure, we can change their rules, but we'd better make sure that the underlying need is still satisfied, because satisfying the underlying need delivers business value. And there, in a nutshell, is my case about bureaucracy and organizational culture:

our goal as Agile practitioners is to understand the underlying need, the business value that must be delivered, and satisfy *that* in the best way possible.

The Rules: we stumble upon a number of rituals conducted by the organization. From where did these practices emerge? Could this be yet another clue?

1   Max Weber, ed. by Guenther Roth and Claus Wittich, *Economy and Society: An Outline of Interpretive Sociology* (Berkeley: University of California Press, 1978), 221.

2   Ibid., 223.

3   Daniel Katz and Robert L. Kahn, *The Social Psychology of Organizations* (New York: Wiley, 1966), p. 222. Cited in Paul S. Adler, "The 'Learning Bureaucracy,'" 61. Katz and Kahn go on to point out deficiencies of bureaucracy as well.

4   Daniel Wren and Arthur Bedeian, *The Evolution of Management Thought*. 6th ed. (Hoboken: John Wiley and Sons, 2009), 233.

5   Ibid., 230.

6   Ibid., 232.

7   The phrase has many connotations and has been translated in various ways. The sense intended here is "dispassionately" or without emotion. It is said to have originated with the historian Tacitus, describing his view of the ideal attitude for a historian.

8   Weber, 975.

9   Ibid., 975.

10 Philip Kotler, *Marketing Management: Analysis, Planning, Implementation, and Control*. 8th ed. (Englewood Cliffs, NJ: Prentice-Hall, 1994), 444.

11 Weber, 339.

12 John Stuart Mill, *Considerations on Representative Government*. 2nd ed. (London: Parker, Son, and Bourn, 1865), 114. Note that Mill was not a fan of bureaucracy. He also said that it tends to devolve into a "pedantocracy."

13 David Graeber, "In Regulation Nation," *Harper's Magazine*, March 2015, 13.

14 As cited in Wikipedia entry for bureaucracy. Reference is to Karl Marx (1970). "3A." *Marx's Critique of Hegel's Philosophy of Right* (1843). Cambridge University Press. Accessed October 12, 2012.

15 Mike Cohn, *Succeeding with Agile: Software Development Using Scrum* (Boston: Addison-Wesley, 2010), 171.

16 Greg Smith and Ahmed Sidky, *Becoming Agile in an Imperfect World* (Greenwich, CT: Manning Publications, 2009), xviii.

17 Schwaber and Sutherland, "The Scrum Guide," 6.

18 Paul S. Adler, "The 'Learning Bureaucracy,'" 64.

"Next time I come here," he said to himself, "I must either bring sweets with me to make them like me or a stick to hit them with."

Franz Kafka, *The Trial*

The more constraints one imposes, the more one frees one's self. And the arbitrariness of the constraint serves only to obtain precision of execution.

Igor Stravinsky, *Poetics of Music*

l = 27 Armadillos

I have an ulterior motive in setting forth the argument of this book. I am a CIO. If we take Schwaber and Sutherland's instruction at face value—that the team is not allowed to listen to anyone except the product owner—then what exactly should *I* be doing to add business value to the enterprise? Throughout the Agile literature, one can see authors struggling to explain the role of management. The point is to empower the team, right? Just as culture and bureaucracy are often viewed as impediments, so too is management. Some authors grudgingly grant management a role—perhaps to remove impediments for the team, perhaps to assemble good teams, perhaps to enforce compliance with those good technical practices that are "not optional." As the management leader responsible for Information Technology, this issue has some urgency for me.

Let me start with another anecdote from my days working with Takeshi and John at Intrax that I think clarifies the challenge.

———————————

When I first joined Intrax as the CIO, I led a very successful transformation of one of our lines of business. We delivered a new system in less than six months that completely repositioned the company in the market and contributed to several years of explosive growth. The system was designed in a true collaboration between the IT organization and the head of that business line.

Of course, other business lines wanted in on that sort of success, so we began a similar initiative for a second part of the organization. Unfortunately, it didn't go nearly as well. The culture of the second business unit was very different—more consensus-oriented. We had arguments between IT and the business about the requirements, but with passionate and strong-willed people on the business side and with the assumption that the business knew what it needed, IT generally gave in.

The launch of the system was a disaster. The CEO, Takeshi, called me into his office and bluntly told me that I had screwed up. I put up a fight: we had done a great job of delivering on the business's requirements, if not right on schedule, then at least closer to it than most projects; we were pretty much on budget; and the system did just what the business had asked for it to do. It was not a problem with IT's execution, it was a problem with the business's requirements and the way it rolled out the new system.

"You are missing the point," Takeshi said. "I have trusted you with an investment in an IT system. Your job is to make sure that I get a good return from my IT investments. I am not getting a good return."

I argued, of course, that he wasn't being fair, that I had no authority over the business unit's management, and I couldn't compel them to change the requirements to something that would be more effective. But ultimately I had to admit that he had a point. Was I just responsible for executing projects, or was I responsible for delivering business value from IT investments?

I preferred to think of myself as a source of business value rather than as an order taker. Many CIOs will say the same thing. Yet even in Agile environments that aim to tear down barriers between the team and the business, our way of thinking about business value still doesn't support that role: the product owner model still has *the business* making decisions about business value, what requirements will deliver that value, and the relative priorities of those requirements. The development team, generally a part of the IT organization, is expected to find solutions but not to define the business need. But IT can deliver on the requirements it is given with exquisite execution— and still not be delivering business value.

Just as the goal of the Agile team is to deliver business value in a particular project area, the goal of the IT organization—if there is to continue to be such a thing—is to deliver value at the macro enterprise level. In an Agile model, particularly in DevOps, the functions that were typically thought of as IT responsibilities are pushed down to teams: software development, testing, infrastructure engineering and provisioning, operations, architecture, and so on. It seems like the only functions left to the broad IT organization are helpdesk operation, procurement, governance, and perhaps hiring. Even these, arguably, belong with the teams: just as developers are now asked to be on call to fix operational problems, perhaps they should similarly be on call to answer helpdesk queries. That too might lead to improvements in code quality and usability!

Of course, IT management should act as servant leaders and remove impediments for the teams that are doing the work. But that just pushes off the fundamental question—exactly what impediments should IT management be helping to remove, and how? In fact, it raises a number of other critical questions about how the Agile teams relate to the IT organization:

- How does IT leadership ensure the delivery of value and participate with the rest of the organization in defining value?
- What role does an IT organization play when teams work directly with the business to create value, when teams self-organize to deliver value, or when a product owner is the only voice that expresses business value decisions?
- If Agile frameworks and DevOps in particular are based on an inclusiveness that extends beyond the boundaries of IT, then what is it that IT is accountable for?
- What can IT provide if not the functional silos of technical experience that it has provided in the past?

The legacy view considers the IT organization a service provider to *the business*. The business formulates business needs and IT delivers on those needs. Its success is measured by customer satisfaction—how well it is serving its business clients—and by its cost-effectiveness. In some aspects, IT is a utility, providing computing power, a network, data storage, helpdesk services, hardware procurement, and commodity software services such as email and office productivity suites. The bulk of IT spending typically goes into these "keeping the lights on" services, and as a result CIOs often report to CFOs. IT is thought of as a category of spending that can be judged on its cost-effectiveness and the service provided by its helpdesk, purchasing, and engineering staff.

Other aspects of the IT function—typically the delivery of software systems—are based on a slightly different model. For these projects, which are typically the high-risk aspects of the IT function, the business develops a set of requirements and gives them to the IT organization. The IT organization then is tasked with delivering on this set of requirements and is judged by its execution ability in terms of cost, schedule, budget, and quality. Some authors have described this role as "IT as an order taker."

Notice how neatly this model fits in with the Waterfall approach to system delivery. The business chooses a set of requirements that it believes will add business value. It communicates them to the IT organization—on paper, of course. The IT organization responds with a plan and estimates of cost and schedule. The business agrees to the plan. The IT organization executes the plan and delivers a product to the business. The business accepts the product and then tries to derive business value from it. This model is the "negotiating a contract" model that the Agile manifesto de-emphasizes.

The initial thrust of Agile approaches was to change the *delivery model*. The requirements were not to be agreed on in advance but refined through interaction between the business and the Agile team. Instead of delivery being in one lump at the end of a long development and testing process, it was to come incrementally in short bursts. The process of system creation was no longer framed as "toss it over the wall"; instead the business and the Agile team worked together to deliver the system.

But the critical assumption that was *not* challenged was that *the business* is responsible for determining the business needs—in other words, for making decisions on what would add business value. Even as the team and the business worked together iteratively and incrementally, it was solely the business that owned the business need. The IT team received frequent feedback from the business on whether it was "meeting the need" and adjusted course accordingly. The essence of the model was still that IT was not part of the business, but again an order taker.

You might say that in the classic Waterfall approach, IT is like the person behind the counter in a fast food restaurant, taking orders and then serving them up. With the shift to an Agile culture, IT became more like the salesman in a fashionable clothing store. Now IT helped the customer decide what he or she really wanted by making suggestions and giving the customer outfits to try on in order to elicit information about what the customer was really looking for. By letting the customer try on outfits, the salesperson got feedback and used the feedback to zero in on what the right product would be. With maybe a bit of tailoring. I've never worked in a fashionable clothing store, but I'm guessing that a good salesperson probably proceeds by first getting the basic outfit agreed upon and then incrementally getting the customer to agree on a shirt, accessories, and shoes. It is an incremental, iterative, and joint process.

But the salesman is still an order taker. It is the customer who makes value decisions.

There is considerable literature on the role of the CIO in an organization and how it is changing. The one point all the authors seem to agree on is that the order-taker mentality just doesn't work. Richard Hunter and George Westerman consider it a classic "value trap"—an approach to IT that *seems* to deliver value but actually does not—to assume that "the business is IT's customer, and the business is always right."[1] The problem, they say, is that business executives don't *want* to be treated like a customer by IT, nor do they want IT to "become aligned" with them—they just want business outcomes.[2] The challenge for the CIO is not to demonstrate that IT has technology expertise or that it completes projects successfully; the CIO's challenge is to show that IT is adding concrete business value to the organization. Not business value in a general sense, but business value as this organization defines it. Even further: IT helps the business determine what it values. "The real value—and the real effort—lies in helping business managers identify how to change the business and then helping them play their roles in implementing those changes."[3] It's also not enough to do the work that *can* create value: IT must ensure that the value is "harvested" through post-implementation reviews and audits.

In his comprehensive study of how to apply Lean thinking across the entire enterprise, Jez Humble points out that the distinction between IT and the business is an outdated one. "In high-performance organizations today," he says, "people who design, build, and run software-based products are an integral part of the business. They are given—and accept—responsibility for customer outcomes."[4] Of course I have to quibble with Jez on his reference to *customer* outcomes; I'd prefer to generalize it to *business* outcomes. But the sense of the IT function as a full participant in business value definition and creation is there. It is hard not to notice that the role that is attributed to the Agile teams is also attributed to the IT organization as a whole.

Peter Weill, who writes about IT portfolio governance, finds in his studies that "the top governance performers were more likely to make business-oriented IT decisions jointly"[5]—that is, IT together with the rest of the organization. While early Agile thinkers arguably were focused on fixing a problem that they saw in software development—development processes that were overly disconnected from end-users and customers or, more generally, the business side of the enterprise—it turns out that the IT function is actually part of the business.

Weill speaks of governance as a way for IT to *learn* about IT value in the enterprise. "Much of the challenge of effective IT governance results from the difficulty of assessing the value of IT,"[6] as he puts it, and understanding the value of IT is difficult because it "cannot always be readily demonstrated through a traditional cash flow analysis. Value results not only from incremental process improvements but also from the ability to respond to competitive pressures . . . it can be difficult to determine in advance how much a new capability or additional information is worth."[7] In other words, business value decisions rest within IT as well as *the business*, and the IT manager is in much the same position that we earlier associated with product owners: lacking a simple, single way to determine value. Governance processes in Weill's view set up *rules* for managing the IT portfolio, and the organization can then learn about IT value by assessing the results of those rules, finding exceptions that work well, and finally institutionalizing the exceptions. "Enterprises in our study," he says, "reported that 50 percent of new systems involved exceptions to their enterprises' normal policies for architecture or investment."[8]

I have argued that business value is not a preordained metric, but something determined by management's interpretation of how to fulfill the organization's ultimate goals. Again, Weill is on the same page. "Top governance performers [in his study] had more differentiated business strategies based on value disciplines such as customer intimacy or product innovation. Governance performance was lower in enterprises pursuing operational excellence. . . . What are the specific desirable behaviors for achieving the enterprise's particular style of operational excellence?"[9] The ultimate goal—let's say maximizing shareholder value—is best translated into a differentiating strategy, which in turn identifies a set of values that can be used to make IT portfolio decisions.

The IT organization, one among many components of the business, is in a unique position when it comes to interpreting the business value of IT investments. As a team of professionals, it understands the state of the art, the emerging ideas and technologies, the best practices for execution. It can see across the enterprise to find synergies across the IT portfolio and to align the portfolio with the business's strategy. It has the accumulated knowledge and wisdom of the organization on what will create enterprise IT value. It understands the possibilities and benefits of technology and what benchmarks to use in assessing performance. It formulates strategies for security and compliance, for use of budgeted funds, and for hiring appropriate technical skill sets. It manages costs and negotiates contracts.

But it remains difficult to see how the autonomous, encapsulated Agile team can participate in this store of knowledge, expertise, and guidance if it only listens to the product owner on questions of business value. Does the prioritization of security-related stories necessarily fall to the product owner? Or is it part of a CIO's vision of how to manage risk across the enterprise? What if the CIO believes that moving from a private datacenter to the public cloud is essential for the long-term IT strategy, and that move requires some work from one of the product teams? Is this for the product owner to decide?

What if it has no net value for this particular project, but it does for the organization as a whole? The product owner may be incentivized to disregard or de-prioritize a valuable unit of work.

Of all of the provocative ideas I have presented in speeches at conferences, the one that has created the most disagreement is the idea that product owners could sometimes be drawn from the IT organization. But if IT really is part of the business, and if decisions are made jointly by the IT organization and business operating units, then why is this an outrageous idea? I think the reaction is based on historical concerns—that user needs were often neglected when projects were too focused on technology. It also echoes the order-taker mentality: IT has no right to be meddling in what is clearly a business concern. But within IT organizations I have seen, there are often employees who understand the company's business operations as thoroughly as anyone, with a broad perspective based on working on projects across the enterprise, and with IT skills and understanding. Perhaps the best hope, if we are committed to the product owner model, is to groom product owners within the IT organization who understand the business well.

To overcome the limitations of the order-taker model, the IT organization must become an interpreter of business value as well as a provider of technical skills, and its interpretation must influence not only the behavior of project teams but also the enterprise's business strategies. And that brings us to our final chapters, which look at how Agile teams can create business value.

The CIO: another shady character enters the picture, claiming that he too has been charged with solving this mystery. Or does he bear some responsibility himself?

1   Richard Hunter and George Westerman, *The Real Business of IT: How CIOs Create and Communicate Business Value* (Boston: Harvard Business Press, 2009), 8.

2   Ibid., 36.

3   Ibid., 139.

4   Jez Humble, Joanne Molesky, and Barry O'Reilly, *Lean Enterprise: How High Performance Organizations Innovate at Scale* (Sebastopol, CA: O'Reilly Media, 2015), 111.

5   Peter Weill and Jeanne W. Ross, *IT Governance: How Top Performers Manage IT Decision Rights for Superior Results* (Boston: Harvard Business School Press, 2004), 135.

6   Ibid., 103.

7   Ibid., 16.

8   Ibid., 17.

9   Ibid., 127.

At last I will devote myself sincerely and without reservation
to the general demolition of my opinions.

René Descartes, *Meditations On First Philosophy*

There is no real direction here, neither lines of power nor cooperation.
Decisions are never really made—at best they manage to emerge,
from a chaos of peeves, whims, hallucinations, and all-around assholery.

Thomas Pynchon, *Gravity's Rainbow*

6

THE CLUE

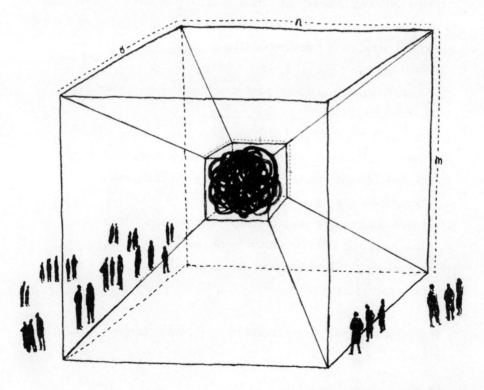

$$a \times b \times c = 784 \; Armadillos$$
$$m \times n \times d = 11,664 \; Armadillos$$

Whe have rounded up the usual suspects: those known to us by three-letter acronyms (ROI, NPV, MVA), suspicious characters and folks lurking in the shadows (product owners, CIOs), mysterious organizations (*the business*, X-teams). Yet the mystery remains. Just what is this *business value* that we want to maximize? Clues abound: organizational culture seems to contain within itself concealed assumptions about value; rules and formal processes seem to have hardened around practices that have been found to deliver it. But the concept still remains elusive, still at large after evading our dragnet, and the urgency seems to be increasing—the Agile team will be judged on it, the product owner has to represent it, and now the CIO as well seems to be accountable for it. And there are only two chapters left. We had better get down to brass tacks.[1]

We've followed Descartes in the demolition of our previous opinions. Now we must reconstruct them on a firmer foundation. Let's quickly summarize our findings so far:

- Business value is not a simple formula somehow known by *the business*.
- It shouldn't be confused with *customer* value or *user* value.
- High-level business goals like "maximizing shareholder value" must still be interpreted into concrete formulations of business value.
- These concrete interpretations of business value are not always explicit and often must be discovered.
- Bureaucracy and culture seem to contain clues to what the organization values.

So now what? How do we create business value through our Agile practice?

The plan is this: we're going to see that business value decisions are strongly influenced by uncertainty about the future, find a definition of business value that takes into account this large role that uncertainty plays, look at some existing techniques that play a role in dealing with this type of uncertainty, and then in the next chapter look at some concrete ideas on how to apply our newfound understanding of business value.

Here we go.

The Waterfall model was based on taking a point-in-time snapshot of the information we know and using it to create a long-term plan that we would adhere to. The Agile insight was that we should change our notion of what features will create business value *over time* as more information becomes available and, in fact, that it can be worth an investment even just to increase learning, thereby reducing uncertainty and opening a space in which innovation can occur. Agile approaches added a time dimension where previously there was none. But the meaning of business value itself was a given: business value is long-term profitability, or it is ROI, or it is something else that is known to the business, and that is good enough.

It's not good enough, though, because in an environment with decentralized decision-making—a team environment, for example—the *criteria* for making decisions have to be pushed to the teams as well. For a team to make good business value decisions, it must understand what constitutes business value and recognize value when it sees it. And this, I have argued, is not as straightforward as it sounds. An organization's understanding of business value changes over time; different organizations have different understandings of business value; calculations of business value must consider uncertainty and the time value of money.

What we must do now is to add two new time dimensions. First, our understanding of business value—the criteria we use for deciding how much value features have—will change over time as the organization changes and learns. Second, the assignment of value to stories will need to consider the time dimension in which that value is harvested. Agile approaches have had a dynamic view of requirements; now we need to add a dynamic view of *how we assign value*.

But before we head into the future, we will have to deal with an elephant that is already in the room. I have glossed over a difficult question in those last paragraphs. Is the Agile team, in fact, responsible for delivering business value? Or is that just the job of a product owner or business representative? Let's see if we can squeeze this elephant back out the door.

––––––––––––––

There is a bit of tension built into the Agile approach. On the one hand, it seems to want to erase the boundary between developers and the business, or at least to minimize it. On the other hand, it seems to impose a sharp distinction between the roles of the developers and those of the people in the business—the business is responsible for deciding what to do and the developers are responsible for deciding how to do it. In a world where technology is so deeply embedded in business, and in an environment of team empowerment, this distinction seems troublesome to me.

Once again, we'll focus on Scrum because it is most specific about this relationship, but keep in mind that analogous questions might apply to any Agile framework that puts business value trade-offs in the hands of *the business*. Now, first, is the product owner part of the Agile team, or separate from it? According to Scrum, the product owner is the OPYCLT—the *Only Person You Can Listen To*. Mike Cohn says, "Part of the product owner's responsibilities is to have a vision of what he or she wishes to build, and convey that vision to the Scrum team."[2] Sounds like the product owner is separate from the Scrum team. ScrumMethodology.com says, "There are three roles in the Scrum method of Agile software development: the product owner, the Scrum master, and the team."[3] Again, sounds separate. Schwaber and Sutherland say, "The Scrum team consists of a product owner, the Development team, and a Scrum master."[4] Interesting . . . the *Scrum* team includes the product owner, but the product owner is separate from the *Development* team. As Roman Pichler writes:

The product owner is required to closely collaborate with the team on an ongoing basis and to guide and direct the team (e.g., by actively managing the product backlog, answering questions when they arise, providing feedback, and signing off work results). In simple terms, the product owner sits in the driver's seat, deciding what should be done and when the software should be shipped.[5]

So the product owner isn't just separate from the team, she is the driver of the work.

Am I the only one who feels a little bit of command-and-control discomfort here? How is the product owner as OPYCLT or as *driver* different from a manager in a hierarchical organization, with the team reporting to her? Perhaps you'll say that the product owner just tells the team what to do, not how to do it. But that is precisely what a good manager *does* anyway. Or perhaps you'll say that the product owner must exist as a special role because someone must have the final say when it comes to prioritization and features. Exactly. A manager.

Is it possible that the product owner role feels like a good idea because it helps fit Agile practice into a command-and-control hierarchy? Or, as I said earlier, because it replicates a loose coupling pattern?

Bringing it back home to business value, and assuming that the product owner is not a part of the Agile team, I will ask this question: who is responsible for delivering business value to the organization—the product owner or the Agile team? If I understand it correctly, the product owner role is explicitly designed to be responsible for business value delivery, since the product owner is the person responsible for making good prioritization decisions in order to maximize that value. The team's responsibility is to deliver those features. How can the team "own" value delivery if their responsibility extends only to delivering features that they neither choose nor oversee the adoption of?

But it doesn't feel right. If this is a team sport, the team should own value delivery. What would it take for that to be the case? For one thing, the team would have to understand what business value means. That is why I wrote this book. But it takes more. The team needs to reach out beyond its boundaries—and beyond its OPYCLT or driver—to touch the organization everywhere from value determination to value harvesting. The team should be an X-team.

Leadership defines the business value context for the enterprise's activities, and the team needs to absorb that context, not ask the product owner to interpret it for them. The team must build relationships that allow it to sense context; it must provoke and observe. The product owner cannot be a bottleneck in this process. The team must learn—but not only about its own internal processes, as the Agile Retrospective concept would imply. Our X-team participates in the organization's attempt to define value in the first place. It conducts value experiments on behalf of the organization to help it learn. With the fast deployment and short feedback cycles established through a Continuous Delivery pipeline, it can experiment with the limits of the organization's tolerance for change, for example. It can find pockets of resistance and pockets of support; it can learn the reasons for resistance and find ways to address them.

A team that "owns" business value delivery must advocate for its own ideas and negotiate its boundaries. It needs skills of persuasion, communication, and emotional intelligence—sales skills. Why shouldn't it also have strategic skills? A team of generalizing specialists—T-shaped people[6]—with IT savvy and the ability to build things themselves. Isn't this a powerful model in a digital services world?

The product owner role can be seen in a different light. Rather than being the authority of business value, making business value trade-offs down to the feature level, the product owner can also be seen as the *visionary* responsible for a product or system, the person who communicates a vision behind a product and steers its development and evolution, the person who establishes a high level business value *context* in which the team generates solutions.

That concept of the product owner's role makes all the more sense now that we have come to think of a business as a type of Complex Adaptive System (CAS). A CAS is a self-organizing system with non-linearities and complex interactions, one in which leadership can influence, but not control, outcomes. An important function of leadership in a CAS is to provide context and incentives that nudge the system toward the desired outcomes. In John Henry Clippinger III's book *The Biology of Business*, a business organization is presented as a Darwinian universe in which leaders and managers can set parameters that determine what ideas and behaviors are most likely to survive: "Management cannot, and need not, have perfect information; the challenge of management is to create the conditions and contexts that select for a range of desired outcomes as in the processes of natural selection."[7] Managers do not have an "Archimedean vantage point," a privileged position above the activity of the organization, but on the contrary, "their role cannot be extricated from, or elevated above, the fray of selection and emergence."[8] In one of the essays in Clippinger's book, Andy Clark reflects on "the nature of control: not as a top-down imposition of respecified form but as the selective application of relatively simple forces to a highly complex self-organizing system."[9]

The role of leadership and management in a CAS is to influence the evolutionary process so that the organization delivers on its goals. Leaders create the language of the organization; they set up incentives and define value in a way that elicits the desired outcomes. They define what is *valued* in order to deliver on the organization's mission. While the ultimate goal of the publicly held corporation might be to increase shareholder value, leadership must interpret that goal into a language and a set of operational values that it believes will deliver on that ultimate goal and at the same time will be effective in influencing the evolution of the CAS. In a private company, a non-profit, or a government agency, leadership may play an even more critical role in making business value concrete. The art of leading a business organization is the art of communicating an interpretation of business value so that teams and individuals in the organization will self-select the behaviors that will produce the mission of the organization. A product owner is simply a manager in the organization who participates with other managers in framing context to influence outcomes.

We can find a similar line of thought in the Agile literature around self-organizing teams. According to Larman and Vodde, for example, "Self-organizing teams do not just happen, they need the right environment. The organization is responsible for supporting the team development by creating the conditions needed for teams to succeed."[10] If we read this as "the organization is responsible for helping the teams succeed by giving them a context in which business value is defined in a way that produces the desired outcomes," then I fully agree. Ancona and Bresman, the proponents of X-teams, believe that "leadership, therefore, must be pushed from the executive level to the operational level, with rapidly flowing dialogue between them. All of this adds up to a move from a tight structure of command-and-control toward a looser organization of coordinate-and-cultivate."[11]

Context, coordination, cultivation, creating conditions for success—all of these suggest a role for management that is about influencing the evolution of the organization in a direction that delivers business value. Even Takeuchi and Nonaka, though they use the old terminology of "control," make a similar point when they say, "Subtle control is also consistent with the self-organizing character of project teams."[12]

In another essay in Clippinger's book, Philip Anderson tellingly frames managerial influence in terms of business value. Considering the business to be a network of teams and individuals, he says that "managers shape networks' behavior by emphasizing indicators that they believe will ultimately lead to long-term profitability."[13] Anderson is saying is that managers influence the way teams think about business value in order to direct the teams toward what that they think will best achieve the organization's desired outcomes. By "indicators" he means the measures of their outputs that demonstrate the value they have added. "Complex Adaptive Systems strive toward better fitness as defined by managers . . . a self-organizing system can only evolve functional behaviors when managers tell it what functionality means."[14]

An example he uses of managerial intervention is particularly interesting in our context: the matter of choosing an appropriate discount rate for calculating NPV. This, as Anderson points out, lets management influence how the organization trades off long-term returns versus short-term profits.

The definition of business value is part of the context management sets to influence behavior. Of the Agile frameworks, XP perhaps comes closest to this notion of a team operating proactively and absorbing influences from the organization in a way that allows it to make decentralized decisions. XP refers to its "whole team" approach; the thought behind its onsite customer role, I believe, is that the team should have everything needed to make independent decisions that are consistent with the company's notion of value. But the literature of XP doesn't dwell on how the team *knows* what is valuable to the rest of the business; the assumption, I think, is that the onsite customer makes the business decisions and determines priorities. As I showed in chapter 1, this identification of business value with customer value is not always appropriate. The XP concept, though, can fit easily with the notion of an X-team operating in a CAS.

But how does leadership know what indicators of business value are the right ones to follow in order to achieve the desired business outcomes? My answer may be obvious by now. They don't know. They have a hypothesis. They can use this hypothesis to influence natural selection in the organization and observe the results. As they see the results, they can alter their course. Natural selection eliminates leaders that do not deliver on shareholder expectations.

Yes, we are ready now for my definition of business value:

Business value is a hypothesis held by the organization's leadership as to what will best accomplish the organization's ultimate goals or desired outcomes.

The organization's goal might be to maximize shareholder value: if so, then leadership's business value hypothesis is about what combination of indicators will lead to long-run increases in shareholder value; it might attach particular value to investments that contribute to its generic competitive strategy. The goal might be to eradicate malaria in Africa, in which case leadership will have a hypothesis about what will best achieve that goal. The goal might be to raise the value of the startup in the next round of financing so that the Series A venture capital investor will have less dilution of its equity.

In any of these cases, leadership's hypothesis is not a single plan of activities, but rather a set of business value indicators that will add up to the desired goal. It will include, perhaps, some revenue generation, perhaps some customer satisfaction, perhaps some risk management, perhaps some brand-building. Each of these is a source of business value that the Agile team will seek to deliver and maximize. The nonprofit that wants to eradicate malaria does not just prioritize activities that eliminate malaria: leadership's current hypothesis may be that a combination of disease-fighting, cost reduction, and marketing to raise awareness about the cause are the sources of business value that will lead to the strongest outcomes. Raising awareness, in that case, is far from being "waste," even though it is not a cure for the disease.

For an example of how this hypothesis-setting works in practice, we can take the case of Netflix. The hypothesis of Netflix's leaders is that retaining new trial customers (specifically not customers who are rejoining the service) to the point where they convert to paying customers will drive shareholder value. The resulting focus on new— that is, future—customers defines value in an actionable way. It creates a context in which Netflix teams know how to make a critical trade-off: the trade-off between adding new features that existing customers are asking for and keeping the product simple to optimize it for future customers. Since retention is a lagging indicator—they won't know for a month whether the trial customer continues as a paying customer—Netflix uses a proxy measure to evaluate value delivered by a new feature: consumption. Customers that watch more videos are receiving more value and expect to receive more value next month, so they are likely to renew. The context created by leadership allows value trade-offs to be made and value-add to be measured.[15]

———————

With this understanding of business value in mind, we can reframe some of our previous concerns. In a CAS where leaders articulate and try out hypotheses to deliver on ultimate goals, the understanding of business value takes on a very dynamic aspect, a time dimension. In contrast, the more superficial notions of business value we explored earlier were based on a static view of what business value means, referring to the future, if at all, by collapsing it into a single value (a risk-adjusted NPV, for example) to represent uncertainty. Let me explain why the earlier approaches make me uncomfortable.

→ **They are fundamentally static.** The notions of business value based on NPV (or ROI in its best interpretation) treat risk as a given and boil it down to a simple metric—for example, a discount rate based on investments with "similar risk." They compress future possibilities and risks into a moment-in-time measure, a "present" value. Our Agile approaches, however, are concerned with managing risk by taking actions over time, rather than by accepting it as a given. The Lean Startup approach, for example, proceeds by using low-cost experiments to reduce risk through validated learning. The traditional method makes sense if investment and prioritization decisions must be made all at once, but it is less relevant if we can stage our decisions over time, where risk measures are constantly changing and where we can decide "at the last responsible moment."

In his textbook on managerial accounting[16], Charles Horngren identifies five techniques used to incorporate risk into investment decisions:

- Insist on a shorter payback period
- Increase the discount rate
- Reduce projected cash flows by a percentage
- Perform a sensitivity analysis
- Use probability weightings of different cash flow scenarios to derive an "expected" outcome

These five techniques, unfortunately, require us to make risk decisions at the present instant, when we have little information that would help us assess possible outcomes. Take the "expected value" approach of the last technique, for example. It has us visualizing several possible outcomes and assigning them probabilities—based on today's information—and using them to calculate a single point-in-time measure that loses all the fidelity, all the detail, of the outcomes that we visualized.

Because our ROI and NPV calculations are so sensitive to revenue and cost projections, and because projections are so uncertain for software features, the calculation of these metrics is "swamped" by the risk factors. But the ROI and NPV calculations have no good way to incorporate new information that we might gain as the future unrolls. They assume that we must make all of our decisions now—in advance—with the information we have. They are the equivalents of the Waterfall in the financial world. In fact, we can often stage investment decisions over time, and we will have options to learn and adjust. An Agile notion of business value would incorporate the time dimension.

→ **They do not pay adequate attention to latent sources of value.** It is one thing to look at a product feature and estimate revenue streams that might be derived from it. But there are sources of value whose impact is much more difficult to value in a simple set of projections. These sources of value can be thought of as assets of the organization, though they might not appear on a balance sheet. They have an economic value—a capacity to influence future cash flows or mission accomplishment, which is quite separate from their accounting value (capitalized software development costs, for example). These include assets like validated learning and information and data. They also include the Enterprise Architecture—the hairball I described in the preface—which is the economic value of the IT capabilities that exist. Perhaps we should even see the organization's bureaucratic rules and corporate culture as an asset, since as I pointed out earlier these are a form of institutional memory.[17] There is a sense in which we need to place an economic (negative) value on the abilities of competitors, since as I pointed out, the latent possibility that a competitor will imitate our feature affects our projected cash flows. The point is that these latent assets may later come to generate a value stream if they are "unlocked" over time.

→ **They rely on magical interpretation.** With NPV, ROI, and other traditional ideas of business value, someone is projecting cash flows or future impacts, and someone is deciding what values to superimpose on these projections for making decisions. What makes this person's interpretation correct? How do we know that this person has the right information available to them? The traditional metrics pretend to be objective but are still the results of an individual's work—an individual who may or may not be fully aligned with the organization's strategy, may have biases or political concerns of their own, and may be incentivized in a way that affects the projections and interpretations.

As we will see in the next chapter, evidence shows that a very large number of ideas do not wind up creating value; people are not very good at projecting how much value will flow from a new idea. Calculating NPV, ROI, or other point-in-time value metrics by having a person predict the cash flows that will result from a new idea is unlikely to lead to good decisions.

I'm simply suggesting that we need to move beyond an idea of business value that simplifies the future into a "present value," that ignores the latent value that may require a catalyst to unlock in the future, and that is mysteriously arrived at and applied by the business, perhaps through a product owner, whose job is separate from that of the Agile team.

Fortunately, there are several wonderful tools available to help us add that fourth dimension to business value: Cost of Delay, Scenario Planning, and Options Thinking.

Cost of Delay was introduced by Donald Reinertsen in his book *Product Development Flow*. In order to demonstrate that organizations should invest in reducing lead times in the product development process, Reinertsen proposes calculating the cost of delaying the introduction of features into production. If the cost of delay exceeds the investment required to reduce that delay, then the investment adds economic value. Since lead time depends on batch size, queue lengths, capacity utilization, variability in demand, and processing time, these are the logical areas in which to invest.

The idea has been extended for use in prioritizing features, for example by Joshua Arnold and Özlem Yüce. In their account of how they used the technique at Maersk Shipping Lines, Arnold and Yüce explain how they used as a comparative metric the Cost of Delay Divided by Duration (CD3); that is, the cost of delaying a feature divided by the effort that would be involved in implementing it. Note how similar in structure this is to ROI: it is essentially a return (negative, in this case) divided by an investment cost. There are two key insights here: first, that we don't need to know everything about what return the feature will give us, just the salient points for comparing it to other features, and second, that there is a complex time dimension involved, since the cost of delay might not vary linearly with the duration of the delay. To capture this non-linearity, Arnold and Yüce talk about different "urgency profiles" that can be used for the calculation.

As with ROI, the magic is in choosing the numerator. How do we calculate a Cost of Delay? In *Product Development Flow*, Reinertsen speaks mainly in terms of profit; his Cost of Delay is essentially the opportunity cost of profits not earned because of the delay. As we have seen in chapters 1 and 2, profit has severe limitations as an economic metric: it is dependent on accounting technique, not necessarily correlated with shareholder value, and not easily applied to many organizations.

The Cost of Delay framework is compelling, but for the reasons I have discussed, I am not sure that we will be able to find a single metric that represents business value or that that metric will be practically computable. For example, consider the value of a security-related feature. If our key metric reflects product adoption, for example, then we will not easily be able to calculate a value for the security feature; if the metric is about profit or shareholder value, then it more easily incorporates the security feature but becomes harder to determine. And for an organization like the Department of Homeland Security, how can we measure the number of deaths it has avoided by stopping terrorists?

On the other hand, Cost of Delay is an elegant and simple way to apply some of the concepts I have introduced. It has the virtue of bringing the time dimension to bear on value decisions.

Scenario Planning was developed by Pierre Wack of the Royal Dutch/ Shell Company to make investment decisions in the 1960s in an environment of extreme geopolitical uncertainty. The idea behind Scenario Planning is to examine the drivers that will impact the organization in the future and, based on those drivers, create several plausible scenarios related to the decision at hand. For each of those scenarios, the organization then formulates a story that explains the scenario, and that can be used to determine how the company would like to see itself in that scenario. Compared to weighted average techniques or any other method of forecasting cash flows for ROI or NPV calculations, the scenario method preserves rich detail about each possible scenario that can be used to better inform decisions. As Peter Schwartz says in *The Art of the Long View: Planning for the Future in an Uncertain World*, "Scenarios are not about predicting the future, rather they are about perceiving futures in the present."[18] As they look to the future, however, they give decision-makers high-resolution images of what that future might look like, rather than the single weighted metric often used in investment planning.

Scenarios are not just about making better decisions in the present instant. Scenario Planning practitioners try to find leading indicators for each scenario. As Schwartz explains, "It is important to know as soon as possible which of several scenarios is closest to the course of history *as it actually unfolds*."[19] In other words, Scenario Planning increases our agility because it keeps all scenarios alive as possibilities, and it trains us to recognize which scenario is occurring and to react accordingly. "Using scenarios," Schwartz says, "is rehearsing the future. You run through the simulated events as if you were already living them. You train yourself to recognize which drama is unfolding. That helps you avoid unpleasant surprises and know how to act."[20] Compared to the simple ROI and NPV calculations, Scenario Planning provides a richer view of the components of business value. It reduces risk and increases agility.

To me, scenarios are a great way to acknowledge and accept uncertainty about future conditions. Rather than pretending that we can account for uncertainty through a point-in-time calculation, we withhold judgment and adjust as the future unrolls.

In "The Options Approach," Avinash Dixit, a Princeton University Economics professor, presents a strong argument that NPV ignores one of the most important sources of business value: real options. "A company with an opportunity to invest is holding something much like a financial call option: it has the right but not the obligation to buy an asset (namely, the entitlement to the stream of profits from the project)."[21] In the financial markets, a call option is an instrument that gives its owner the option to buy a security at a given *strike price* on a given *exercise date*. Today, for example, Apple's common stock is trading at $128.70. I can buy a call option that would allow me to buy Apple stock about one month from now at a strike price of $140.00 for about $0.41 (plus fees). One month from now, if Apple is trading at more than $140.00, then I would exercise my option, by the share for $140.00, and resell it for whatever the shares were trading at, thereby making a profit. If on that date the share price is less than $140.00, I would just let my option lapse, thereby losing $0.41. By buying the call option now, I am risking a small amount of money to have the option of making money later if market conditions are favorable.

Dixit is pointing out that just as the financial option has value, $0.41 per share in this case, so the company's option to make an investment has value. When the company decides to invest, it is effectively "exercising" the option, at which point the option itself loses its value. So the real question is how to exercise options optimally:

By deciding to go ahead with an expenditure, the company gives up the possibility of waiting for new information that might affect the desirability or timing of the investment. Thus the simple NPV rule needs to be modified. Instead of just being positive, the present value of the expected stream of cash from a project must exceed the cost of the project by an amount equal to the value of keeping the investment option alive.[22]

Simple NPV, as we have said, makes the mistake of assuming that the investment needs to be made right now. For investments that could be delayed until more information is available, NPV must also consider the value of the *option* to invest. From an Agile perspective, this is a fascinating point. On the one hand, it is just another way of expressing the Lean principle, "Decide at the last responsible moment." On the other hand, it suggests that if we use the Cost of Delay or any of the "instantaneous" measures like ROI or NPV to prioritize features, we should also take into consideration the *value* of delay. As a general rule, that value rises the more uncertainty there is: "The greater the uncertainty over the potential profitability of the investment, the greater the value of the opportunity and the greater the incentive to wait and keep the opportunity alive rather than exercise it by investing at once."[23]

One consequence of options thinking is that investments that create an option have more value than NPV would suggest. Investments that give the company the possibility—but not the obligation—to make more investments later have more value than just their cash flows. Each feature that the company adds to its software product is more valuable to the extent that it gives the company a basis for adding additional features, or for adding additional revenue streams if it chooses to invest further. At the same time, it hedges against risk because the company can wait and see if the additional features or revenue streams will still have value.

Another consequence is that companies may choose to invest in activities that yield information that will give it opportunities in the future. R&D is typically an example of such an activity. The business intelligence dashboard mentioned earlier can be thought of as creating option values.

Real options, in other words, are a way of valuing agility. Each feature we build has value not only for its direct cash flows, but for the increase in the organization's agility: its ability to change course or build on the feature as appropriate in the future. It is strange that we in the Agile community focus so much on ROI—which ignores option values—when we attach so much value to option creation!

These techniques—Cost of Delay, Scenario Planning, and Options Thinking—allow us to consider the future without losing as much fidelity as we do by boiling it down to a point estimate. A point estimate makes sense when all our critical decisions must be made in advance, but Agile approaches let us put off decisions until the last responsible moment and to change decisions over time. To capture this benefit, we need to use these more flexible tools.

The Clue: a breakthrough in the case. Following the CIO, we stumble upon the guardians of the solution: a group called The System, or the CAS. We observe them creating the rituals and norms, looking both forward and backward in time, defining value as we watch. The case solved, we make a pact. The Agile team will be responsible as a whole for delivering business value. Then we give them a set of tools—Cost of Delay, Scenario Planning, and Optionality—to help them navigate the fourth dimension within which business value unfolds.

1   Like business value, no one seems to actually know what this expression means. See the excellent entry in "The Phrase Finder" at http://www.phrases.org.uk/meanings/get-down-to-brass-tacks.html.

2   Mike Cohn, "Topics in Scrum: Product Owner," https://www.mountaingoatsoftware.com/agile/scrum/product-owner.

3   http://scrummethodology.com/scrum-product-owner/

4   Schwaber and Sutherland, "The Scrum Guide," 5.

5   Roman Pichler, "Being an Effective Product Owner," https://www.scrumalliance.org/community/articles/2007/april/being-an-effective-product-owner.

6   The term is attributed to IDEO CEO Tim Brown, who explains that "the vertical stroke of the 'T' is a depth of skill that allows them to contribute to the creative process . . . The horizontal stroke of the 'T' is the disposition for collaboration across disciplines . . . T-shaped people have both depth and breadth in their skills." Morten T. Hansen, "IDEO CEO Tim Brown: T-Shaped Stars: The Backbone of IDEO's Collaborative Culture" ChiefExecutive.Net. T-shaped people are increasingly in demand for Agile teams.

7   John Henry Clippinger III, ed., *The Biology of Business: Decoding the Natural Laws of Enterprise* (San Francisco: Jossey-Bass Publishers, 1999), 3.

8   Ibid., 23.

9   Andy Clark, "Leadership and Influence: The Manager as Coach, Nanny, and Artificial DNA" in Clippinger, 55.

10 Larman and Vodde, *Scaling Lean and Agile Development*, 194. Remember them from chapter 1?

11 Ancona and Bresman, *X-Teams*, loc. 1501.

12 Hirotaka Takeuchi and Ikujiro Nonaka, "The New Product Development Game," 145.

13 Philip Anderson, "Seven Levers for Guiding the Evolving Enterprise" in Clippinger, 132.

14 Ibid., 133.

15 According to Adrian Cockcroft, a former Netflix Cloud Architect, in a private communication.

16 Charles T. Horngren, George Foster, and Srikant M. Datar, *Cost Accounting: A Managerial Emphasis* (Englewood Cliffs, NJ: Prentice-Hall, 1994), 736-737.

17 I'm cheating a bit here: we can all agree that these "assets" sometimes have a negative value as well.

18 Peter Schwartz, *The Art of the Long View: Planning for the Future in an Uncertain World* (New York: Currency Doubleday, 1991), 36.

19 Ibid., 246. Emphasis is mine.

20 Ibid., 192.

21 Avinash K. Dixit and Robert S. Pindyck, "The Options Approach to Capital Investment" in *Harvard Business Review*. May–June 1995, 106.

22 Ibid., 106.

23 Ibid., 110.

The greatest challenge to any thinker is stating the problem in a way that will allow a solution.

Bertrand Russell (perhaps apocryphal)

It is hard enough to remember my opinions, without also remembering my reasons for them!

Friedrich Nietzsche, source unknown

7

THE DELIVERY

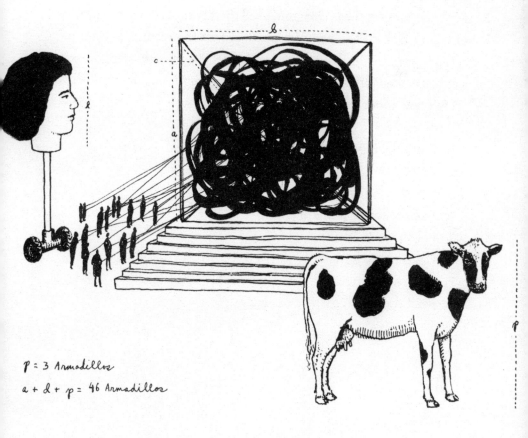

$p = 3$ Armadillos

$a + d + p = 46$ Armadillos

I have never liked books that are very prescriptive. Sometimes the prescriptions strike me as obvious; sometimes they strike me as based on very limited data ("I did it this way a few times and it worked"); sometimes they are based on case studies that reflect a selection bias ("It works for all my clients, but my clients happen to be the ones who read my book and thought it applied well to them"); sometimes they feel like the author saw every example through the lens of his theory and therefore proceeds with a sort of confirmation bias. Prescriptions seem especially inappropriate in a book that promotes experimentation and makes the case that even the very definition of business value varies from organization to organization.

So in this chapter what I'd like to do is to propose some experiments that seem in line with what we've discovered about business value. I've tried some of these ideas and liked them; others I'm in the process of trying, but it's too early for me to report the results. If you can't actually try out the proposals, consider them thought experiments and see if they influence the way you think about business value.

Let's talk about eight "plays":

1. Build the pipeline
2. Extend the pipeline
3. Search for gems in the waste
4. Explore the fourth dimension
5. Polish the hairball
6. Govern wisely
7. Change partners
8. Feed the CIO

1. BUILD THE PIPELINE

I lied! I said I wasn't going to be prescriptive, but I don't see how you can take a disciplined approach to business value delivery without adopting a DevOps culture and a Continuous Delivery pipeline.

The first need when adopting a business value-centric Agile approach is to set up a production process that allows for rapid, informative, and reliable feedback cycles. The best way we know to do this today is through the complex of practices we call DevOps. DevOps combines a highly automated, repeatable, and testable Continuous Delivery framework with a unified team that extends through the full lifecycle of delivering capabilities to production. With a DevOps approach, we can take small batches of requirements quickly from concept to deployment, monitor the results, and use what we learn to change the product and the process.

At the heart of DevOps is a machine—a Continuous Delivery pipeline. It is illuminating (and amusing) to think of this pipeline in light of my comments on bureaucracy in chapter 4. The pipeline is an artifact that incorporates our best understanding of how to create value through software delivery: it is a set of rules and standardized processes, and it requires conformance. As Poppendieck and Cohn have said, good technical practices are not optional! Among those practices that today are found in most CD pipelines are the rule that a broken build should be fixed immediately and the idea that infrastructure should be immutable, that virtual machines should be "cattle" rather than "pets."[1] The pipeline is an automated bureaucracy: it applies its rules in a rigorous, unemotional way, *sine ira et studio*. That does not mean that the software development process is unemotional; it means that the tools are unemotional and the passion is brought to the process by the people.

The testing process plays a special role in this DevOps bureaucracy.

Tests are simply rules in version control. They impose a compliance requirement on the code. Compliance requirements themselves can even be delivered as tests; a system is compliant if it passes those tests. A system is secure by definition if it passes the security tests; it is accessible if the tests say it is.

The pipeline implements the "gating" mechanisms or change control that the organization requires. It is auditable. It provides transparency that supports management oversight. Automated compliance tests together with auditability and transparency give us an elegant way to handle bureaucratic requirements—you might think of it as a kind of "bureaucracy-driven development" (BuDD?) process. But the CD pipeline is a learning bureaucracy. It is tuned through retrospectives and continuous improvement; at any given moment it embodies the participants' beliefs about the best way to generate value through software delivery. Tests may change, as they are simply artifacts in version control.

The DevOps stance is humble; it respects and learns from the values embedded in the organization. At the same time it is active; rather than passively receiving business value parameters from the business, it can deliver what appears to be value, assess the results, and thereby contribute to the organization's understanding of business value. And through humane practices, it supports the transition from a world where a "project" is tossed over the wall to a team which then kills itself to deliver, to a world where the team is a core part of the business's evolution and self-discovery. By fitting into the organization's processes, capturing practices in rules that evolve, absorbing indicators of value from the culture, and proactively engaging in the definition of value, DevOps is the ideal way to address those aspects of business value I presented in the first part of this book.

Try this:

➔ **Create a highly automated, reliable, Continuous Delivery pipeline that allows for rapid feedback all the way through to the operation of features in production.** Use this feedback to tweak the product so that it delivers on the organization's current understanding of business value. Within the pipeline, encapsulate best-known practices for delivery; continuously improve the pipeline in parallel with continuously improving the product.

➔ **Assemble teams with cross-functional skill sets.** Since the team will have responsibility for delivering business value, not just for developing features, you will need a wide range of skills. Experiment with which skills you include on the team and which skills are available to it as external resources.

➔ **Accept that the delivery pipeline is governed by rules and constraints.** This is the part of the value delivery process that is more about institutional memory encoded in rules and less about self-organization and autonomy. It is "governed" by an organization of functional experts. This also helps make the process auditable; it aligns delivery with oversight and produces artifacts that deliver bureaucratic or compliance value.

➔ **Move toward single-piece flow, or at least very small batches of requirements.** This reduces lead time, leading to faster feedback and earlier value delivery. More importantly, it reduces risk and therefore bureaucratic waste: the riskier a project is perceived to be, the more bureaucratic overhead will be added to it.

2. EXTEND THE PIPELINE

Let's think of our product development process as a black box. In the traditional way of thinking, the inputs to that black box are requirements and constraints; the outputs are deployed capabilities.

The DevOps concept has extended this idea further into the value stream. The inputs are business needs; many constraints are absorbed into the black box since Operations, Security, and others participate in the value creation process rather than constraining delivery. The output is no longer a deployed feature but a feature operating in production, and an additional feedback loop is added.

But the fundamental ideas of DevOps need to be pushed further to accommodate the framework I am proposing. Feedback from a feature operating in production is not enough: we need feedback on how the organization is harvesting value from the new feature after it is placed in production. The team needs to know—and influence—outcomes if it is to know what incremental business value it has created. While Scrum implies that the product owner oversees value harvesting, the DevOps principles can be extended to include the "whole team."

There is also a new feedback loop. At a larger scale, the enterprise has a hypothesis about what creates ultimate value—shareholder value, mission value, owner satisfaction, or whatever it might be. That hypothesis—the definition of business value, in my formulation—is also subject to continuous improvement through feedback. This definition of business value is applied in the incentives, visions, criteria for success, and strategies that leaders use to mold the evolution of the organization as a complex adaptive system.

I have worked with teams that brought together different types of reporting from across the enterprise to gain a more complete view of business value harvesting: production monitoring, audit logging and reporting, development pipeline health and status monitoring, and the organization's business intelligence system. The last of these is used to provide indicators of business value, and the dashboards are integrated and accessible to the team. They have also pulled datasets of business indicators into statistical analysis tools to get deeper insight into the mechanics of value generation and use the results to inform the development pipeline (for example, what new features might eliminate unpleasant outliers in business result?). This is just a start—we need new ways for the teams to extend their reach into the harvesting of business value.

Try this:

→ **Extend the DevOps pipeline to include value harvesting and wider stakeholder involvement.** Since the team is responsible for delivering business value, rather than for delivering and operating features, it must track its product all the way through to its business impact. The DevOps team may need additional skills to incorporate this activity; it certainly needs new ways to monitor business results, perhaps through business reports and analytic tools, perhaps by engaging with users in an operational setting when they are using the system.

→ **Implement BI-Driven Development (BiDD?).** That's BI in the sense of Business Intelligence. In the same sense that we do Test-Driven Development, we can set up dashboards in the BI or reporting system that will measure business results even before we start writing our code. This allows us to focus our efforts, continuously deliver features to production, observe their effect on the business value proxy metrics, and decide when we have delivered enough to move on to other sets of impacts.

→ **Widen the team's inputs.** DevOps is already an inclusive approach. The team can seek input from many areas of the organization, in the true spirit of the X-team. It might be useful to know how the team's ideas fit in with Marketing's brand strategy, with the help-desk's experience in what worsens its workload, and with timing issues that the CFO might be worried about for revenue and expense recognition. It might need information about compliance requirements so that it can produce the most helpful audit artifacts. Notice that in theory, any of these business-functional areas can be placed in an inhumane position by a development process that ignores their concerns.

→ **Consider the Spine Model.**[2] Developed by Kevin Trethewey and Danie Roux, the Spine Model interprets human systems in terms of a series of interacting layers: Needs, Values, Principles, Practices, and Tools. The Spine Model fosters conversations to ensure the alignment of these layers. Especially interesting in our context is Tretheway and Roux's idea of feedback cycles that operate among these layers, with faster feedback informing the lower layers (Practices and Tools) and slower feedback informing the upper layers (Needs, Values, and Principles), with these latter three being the layers that leadership uses to set context for the organization's activities.

3. SEARCH FOR GEMS IN THE WASTE

You might want to be careful when you are deciding what is valuable and what is waste. When we limit our view of business value to *customer value* or *user value*, we see anything that does not add to user satisfaction as waste. Artifacts that can be used for oversight and governance appear to be waste. Process complexity around hiring, procurement, time tracking, and so on seem like waste. Security compliance teeters on the brink of being waste. But business value is really a much broader concept.

As I pointed out in chapters 3 and 4, some of this "waste" is actually adding value. In a government agency with responsibility to the public and to Congress, proving that you are following the rules and using public resources effectively is actually a value-adding activity. It affects your ability to continue to receive funding. Obtaining a clean audit report has value for companies, as does complying with HIPAA, Sarbanes-Oxley, or any other applicable frameworks. Compliance with PCI allows a company to accept credit card payments—surely a value-add. Legal review can reduce risk. Complying with Marketing's branding guidelines strengthens the brand and thereby provides business value. The real question with these sorts of requirements— as with all others—is whether they can be met in a leaner way.

Try this:

➔ **Look more deeply at requirements and constraints that seem to impose waste—bureaucracy, compliance, accounting.** What is their purpose? Do they add business value (not customer value)? If so, how? Find the leanest way to deliver this value; look for minimum viable solutions.

➔ **Negotiate administrative and compliance overhead.** There is usually someone to whom we must answer for compliance. What does that person really need? Can policies be changed or waivers be granted? Make sure they understand the cost of compliance. As Hyman Rickover, a former admiral of the US Navy once said, "If you are going to sin, sin against God, not the bureaucracy. God will forgive you but the bureaucracy won't." Instead of sinning, I suggest negotiating.

➔ **Use Impact Mapping to find those things that impact value but are not necessarily product features.** Impact Mapping is a technique described by Gojko Adzic in his book *Impact Mapping*.[3] It brings together a group of stakeholders from around the organization to brainstorm a mind map that identifies goals, actors who can help accomplish those goals, metrics that can measure them, and activities that can contribute to them. His concept is flexible enough to apply to enterprise IT needs and even to compliance and oversight needs.

➔ **Co-opt compliance folks by bringing them in early and making them part of the team.** This bit of jiu-jitsu can make them responsible for finding Lean, effective ways to comply with their own requirements, as well as giving them incentives to do so.

→ **Co-opt them even more.** Have them be hands-on in solving compliance issues. Where I inherited a QA organization that considered itself responsible for practicing a "gotcha" kind of quality review, I made them responsible for ensuring that projects "built quality in" so that their review would never catch any problems. Where over-documentation was a problem, I told the document reviewers they could only approve documents that were as terse as possible, and in some cases I had them write the documentation themselves.

4. EXPLORE THE FOURTH DIMENSION

The preceding chapter introduced techniques that can be used to model business value as something that unrolls over a time dimension. We can use these techniques to "make decisions at the last responsible moment"—to attach a value to keeping options open until more information becomes available.

Try this:

→ **Use Scenario Planning to model possible future states.** This can be especially useful in competitive game situations ("What if the competitor undercuts our price?") or in situations that are subject to highly unpredictable external forces—geopolitics, for example. Scenarios can be used with Impact Mapping or with Options Thinking to develop possible responses to different scenarios, or they can be used to recognize scenarios as they develop so that plans can be adjusted.

→ **Find opportunities to create valuable real options.** Building feature A may have value in itself, but it may also have value because it facilitates building feature B. A flexible architecture has extra value because it facilitates creating future value in a cost-effective manner. This technique must be used carefully, though. In Agile practice, we generally try to build the simplest design that will meet current needs and avoid over-designing for uncertain future requirements. I'm not suggesting that we change that. I'm talking about building in flexibility through simple design.

→ **Use Cost of Delay as part of the prioritization process.** Cost of Delay begins to capture the time dimension, as we have seen. It considers time-sensitive aspects of delivery in a way that is difficult for the traditional financial metrics. When used with a weighting, as in CD3, it even captures the intent behind ROI—a ratio of return (negative in this case) to effort.

→ **Consider stories optional.** When creating a backlog, rather than just imagining a single way of building the system, have the team or the product owner brainstorm a number of optional stories to choose from as the future unfolds. To stimulate creative options thinking, the team can even generate four to five times as many stories as it will actually use.[4]

5. POLISH THE HAIRBALL

Back in the preface I referred to the organization's Enterprise Architecture (EA) as a hairball, an ugly agglomeration of all of the legacy loose ends that the Information Technology infrastructure has picked up. In many organizations, Enterprise Architecture has connotations of bureaucracy, constraints, or over-documentation. That is not the Enterprise Architecture I am talking about here. I am simply referring, as an abstraction, to all of the systems, applications, infrastructure, and other miscellaneous junk that we have in place to run the organization. Face it: we all have a legacy architecture.

What we need to do is find and tune the latent possibilities that lurk in assets the organization has built. The hairball is one of these assets. Other assets include learning that the organization has accumulated (possibly even "recorded" in its bureaucracy and corporate culture), data that the organization has in its databases, and real options that it has created. All of these assets should be maintained and cared for.

Like it or not, all of the Agile teams are adding to the EA hairball. This seems overlooked in the Agile literature: we think of autonomous-ish teams building products defined by product owners, but we rarely reference the impact on the enterprise hairball and its implications. New features produced by teams are the sticky stuff—chewing gum, duct tape, wax, technical debt—that get attached bit by bit to the hairball.

But that hairball is a corporate asset with economic value. In *Enterprise Architecture as Strategy*, MIT researcher Jeanne W. Ross and her coauthors argue that companies need to build and evolve what they call a Foundation for Execution: a combination of an operating model, an Enterprise Architecture that aligns with it, and an IT engagement model. The Enterprise Architecture, in their model, is the organizing logic of the company's operating model—essentially the company's operating capabilities translated into IT.[5] By creating this unified foundation for execution, companies "have made IT an asset rather than a liability and have created a foundation for business agility."[6] Business value flows directly from creating the optimal Foundation for Execution.[7]

Ross suggests that building such an Enterprise Architecture does not necessarily require a lot of work, but rather that it can be built one project at a time.[8] I would amend that to say one feature at a time in our new world approaching single-piece flow. Interestingly, she relates this to scenarios: "A strong foundation for execution," she says, "prepares the company for unknown future customer demands."[9]

The EA not only supports the current operating model but also has latent possibilities for enabling future operations,[10] and it has costs that exert a drag on future value creation. The hairball must be managed.

Try this:

→ **Treat the Enterprise Architecture as an asset with economic business value, both realized and latent.** The latent value is something like a real option. A set of stakeholders outside of the Agile team—probably within the IT organization—has the responsibility for grooming or polishing this hairball. The channel of influence from those stakeholders to the team must be opened and optimized: bureaucratic controls are not the leanest way to achieve this.

→ **Instead of thinking of business value as a return from an investment in a project, focus on the overall corporate IT system as an asset.** That asset has value just as other assets have value on the corporation's balance sheet (although we are talking here about "economic" value, not accounting value, so the balance sheet doesn't quite capture it). The asset has some value now and will have incremental value when the new set of features is added. But the amount of that new value depends not only on the feature itself but also the synergies it shares with the rest of the asset, the new options that the asset gives the organization, and so on.

→ **Consider the value of information and data, which is often only indirectly related to revenues, costs, and mission accomplishment.** Douglas Hubbard provides an excellent way to think through this value and make decisions on whether to invest in obtaining information.[11] He describes information as a way to reduce risk when making a decision: the value of the information is how much risk is removed, or more technically the reduction in the Expected Loss if the decision turns out to be wrong. Hubbard's approach is perhaps too limited, since it only considers the value of information used to support decisions. It does not directly address, for example, the value of information that might persuade an investor to invest in your company, or the value for a government agency in being able to report results to Congress.

→ **Use Innovation Accounting to capture the latent value of learning in a context of high uncertainty around innovation.** Eric Ries's framework provides a way to think about risk reduction and opportunity identification in a rigorous, actionable way. The idea of validating a value hypothesis and a growth hypothesis is framed in product-centric terms, but it applies to enterprises as well. As Humble points out, two-thirds of good ideas do not improve the metric they were intended to improve[12]—presumably, they generate no value. When we attach a projected business value to a feature, the best way to confirm that it exists is to try inexpensive experiments that test the hypothesis. With each pass through the feedback cycle, we have generated knowledge that has latent business value. One small modification to Ries: in enterprise contexts, the product often already exists, or at least the hairball already exists, so the idea of an MVP doesn't quite capture what we are doing. It is more to the point to think in terms of small incremental modifications to the hairball that will test a hypothesis.

6. GOVERN WISELY

Governance frameworks are typically set up around large batches of requirements: a common approach is to group a set of business needs or requirements into a *program* or *project*, then make an investment decision on whether to proceed with it or not. Even in more contemporary practice, where we make high-level investment decisions on "investment themes" or other such abstract concepts, then leave them to be detailed into requirements later, the investment decision is based on something that becomes a large batch of requirements.

On the other hand, our practices now allow us to approach single-piece flow: as I described it in the preface, Continuous Delivery creates almost a "calculus" of smaller and smaller batches, with correspondingly smaller risk as individual units. I wonder if we don't have something of an impedance mismatch here, where we govern in large units and execute in small units. Of course it would be inefficient for high-level executives to make investment decisions for each small requirement, but it also seems like we are missing something by not taking advantage of small batch flow to optimize investment decisions. What if we could find a cost- and time-effective way to reduce the batch size in which investments are overseen?

This mismatch can be costly. If an investment decision is made to finance a set of themes, capabilities, epics, or what have you, but later it is discovered that some parts of that theme have a questionable business case, then there is no natural way to eliminate the scope and return money to the organization. It can be done, of course, but it is rarely incentivized: the product owner, I think, tends to view capacity as fixed and prioritize requirements within that fixed capacity. As long as there are features that will yield some return, the product owner is liable to continue with them.

Another way to look at this is that the governance body, which makes periodic decisions to fund large batches of requirements, is not able to take advantage of what is learned during the development process to adjust those investment decisions. The investment decision is fixed; the product owner or other decision-maker then works with that investment and takes advantage of learnings to make the best use possible of the investment within the scope of the program. We learn on the scale of single requirements, but make investment decisions on the scale of programs or investment themes—thus the impedance mismatch.

There is an analogous mismatch in the way program investment decisions fit with normal budgeting cycles, which typically are done on a regular annual basis. Programs, on the other hand, rarely align with these annual cycles. Nor do small batches of features.

How to fix this? Honestly, I don't know. I'm working on it now. But here are some ideas for things we might all test.

Try this:

→ **Invest in impacts or outcomes rather than programs.** In a way, this is equivalent to having leadership set the incentives and measure that influence the evolution of the system. A business case can be made up front that there is a good probability that the impact can be realized, but then the managers, product owners, and development team can adjust what they are creating as they learn how to best achieve those outcomes. Impact Mapping, described above, is a useful tool for exploring potential impacts.

→ **Incentivize teams to fail productively.** If a team comes to believe that the goal cannot be accomplished or accomplished cost effectively, it should return to the governance body to have its project eliminated or changed.

→ **Shorten the investment review cycle through continuous transparency.** If we can reduce the costs of the investment review process, then we can cost-effectively build in feedback from oversight as a type of feedback processed by the team.

→ **Move Beyond Budgeting.** Governance through an annual cycle is arbitrary: in an ideal world, we would make decisions about the allocation of financial resources throughout the year as information becomes available. Bjarte Bogsnes, a pioneer in the Beyond Budgeting movement, has described how he implemented these techniques at the Norwegian energy company Statoil and how other companies have used similar approaches to move away from annual budget cycles to more adaptive approaches.[13] The techniques they have used may help us solve the mismatch between the budgeting cycle and the shrinking requirement batch size.

➔ **Crowdsource internally with an open source-like model.** This is an approach that I have been experimenting with, though it may sound strange. What if the components of the enterprise's IT architecture were freely shared across the organization in a sort of open source model? A team could accept contributions from across the enterprise. This could include teams that the CIO creates to handle particular technical issues; teams building related products that need to interface with the team's product; individuals who happen to have technical expertise; and individuals who can contribute business knowledge, design skills, feedback, and documentation. A team could maintain control over what commits would be accepted into the system. Teams working on other projects could fork code and use it or create shared code. As a governance model, this helps create reusable code and might improve the allocation of resources.

7. CHANGE PARTNERS

I argued in the last chapter that the relationship between the team and its product owner or onsite customer does not quite fit with the spirit of Agile approaches. This is part of a broader question about how the Agile team relates to the entire enterprise of which it is a part. Is it "loosely coupled" through a product owner or other business representatives? Does it fit into an organizational hierarchy—say, an IT department? Is it "managed" somehow in a chain of command? Does the organization need to change in order to accommodate the Agile team?

When the product owner is the OPYCLT, the IT organization is forced to influence the behavior of teams by creating policies, rules, and constraints that the OPYCLT cannot bypass. This is not generally the optimal solution for the enterprise.

Here are some ways to think differently about the relationship between the product owner and the team.

Try this:

➔ **Do *not* insist—fist pounding—on immediate, broad, cultural change.** Instead, learn from the culture and move it incrementally by provoking and observing.

➔ **Consider Larman and Vodde's LeSS (Large-Scale Scrum) framework.** It has much in common with the ideas I present here, although I think its treatment of business value doesn't go deep enough. Their idea of Feature Teams moves in the direction of giving teams ownership of the entire process of value delivery; their inclusion of the X-team concept and their idea that the product owner is a high-level visionary rather than a day-to-day feature juggler sidesteps the "loose coupling" problem I have mentioned and gives the team responsibility for making business value decisions.

➔ **Give the team ownership of the entire value delivery chain.** Let the team find its own resources, going beyond the OPYCLT to obtain information wherever it believes it to lie and asking questions about anything relevant: for example, security and enterprise architecture, properties of systems it will be interfacing with, marketing goals, financial considerations, and so on. Involve the team in implementation of business process changes and give them the means to track value harvesting.

➔ **Make the team members ambassadors of their own ideas, advocating for the team's positions and negotiating outcomes.** Software engineering is now about sales as well as design patterns and buffer overflows. In *To Sell is Human*, Daniel Pink reports that his studies show that people who are in non-sales positions actually spend about 40 percent of their time at work engaged in sales activities: persuading, convincing, and influencing people.[14] The emphasis on sales activities is especially prevalent in entrepreneurial settings—which Agile development activities actually are—and is certainly to be expected in an environment with flattened hierarchies and deeply interconnected teams—in other words, a Complex Adaptive System.

→ **Assemble true full-stack teams: teams that can sell, negotiate, drive change, analyze business processes, and evaluate financial outcomes.** And do some pretty serious system performance optimization.

→ **Consider drawing product owner-like people from the IT organization.** I know this seems like a step backwards, but circumstances have changed. Technology is increasingly central to business operations, the product owner role requires full-time dedication, and the IT organization knows what latent value might lurk in the hairball. Should the product owner be someone familiar with today's business process or someone with an interest in driving change?

→ **Think of the product owner or onsite customer as an information source, a team member, a helper—anything but an OPYCLT.** According to Scrum, there are only three roles: product owner, Scrum master, and team member. Perhaps there should only be *one* role: team member. Business representatives, in this way of thinking, are simply team members who contribute business domain expertise.

→ **Have the product owner report to the team.** The team recruits a product owner and onsite customers to fill a need it has—making business decisions and getting feedback that is informed by a deep familiarity with the business. The team can determine who can best fulfill the role, give feedback to the product owner on his performance in the role, and replace the product owner and onsite customers as it sees fit. The team should decide what information it needs from the business in order to best make value trade-offs, and arrange to get business representatives who can provide the information that the team needs. The team can consult, negotiate for the participation of, observe different business specialists as they decide they need. Onsite customers are a great help; the team should make sure they get some and that they are the right people.

→ **Let the teams self-organize product ownership.** A team can possess cross-functional skill sets, including business familiarity. Note that in this vision, IT folks can possess crossover skills—including familiarity with the business—and business folks can possess IT familiarity. The team can decide its own process for making prioritization decisions, perhaps by choosing one among their number as the "business expert" who owns business trade-offs.

8. FEED THE CIO

The poor CIO still has not found his place in the Agile world. But I think the discussion in this book can help us serve up a tasty role for the CIO, or at least some food for thought. The CIO is the leader within the CAS whose primary focus is to help steer the business in its use of information technology. The CIO's impact is not just on the IT organization, but on the entire organization's use of technology: while the CIO is seen as the leader of the IT silo of the organization in a traditional, hierarchical structure, in a CAS, influences do not just come down through a siloed management chain. On the contrary, the CIO's influence extends throughout the entire organization through his management of IT assets, strategy, and value creation. The CIO applies information technology knowledge to help steer the entire organization.

Try thinking of the CIO's role in these ways:

→ **First, the CIO is one of the leaders who sets the business value context, the incentives and success criteria that guide the CAS in its evolution.** The CIO brings a focus and an expertise in matters of information and technology to the role, and thus she acts as a partner to those leaders whose focuses and expertise are finance, marketing, and so on. A large part of this role is a visionary one: the CIO must look forward and help drive the organization toward a future state that is effective given changing technology. The CIO is especially critical in fine-tuning the organization's trade-offs around risk.

→ **Second, the CIO is an architect, or maybe a symphonic conductor.** He sets up teams, charges them with various missions, and provides a framework in which these teams can interact and influence one another. He establishes metrics and incentives that drive the teams toward the outcomes he envisions, without direct command-and-control involvement. While teams may collaborate, evolve architectures, and influence direction, the CIO facilitates these exchanges, sets context and direction, and generates portfolio work items that move the architecture in the right directions. In this role, the CIO also manages the supply chain—the acquisition of tools and licenses to support the team initiatives, the manager of finances and budget. He steers teams toward an architecture that will provide cohesion between the work of different teams.

→ **Third, the CIO owns the enterprise architecture hairball, which as we have seen is a representation of business capability.** As Ross et al. say, "The CIO is a key driver—in most companies the CIO is *the* key driver—of enterprise architecture benefits."[15] I like that framing—the driver of the benefits. The CIO cares for the asset, the enterprise architecture as strategy, to make sure it delivers benefits. She polishes the hairball. The architectural portfolio can have cost synergies (think reusable components), option value, scenario provisions, and usability advantages.

→ **Fourth, the CIO is an impediment remover, servant leader, and an optimizer of the black box of development.** Value creation, as we have seen, requires both choosing the right increments to develop and fine-tuning the development process. Improvements to the development process are highly leveraged: a small improvement in the feedback cycle time affects every one of the increments that flow through the process. The CIO lends authority to the team as needed to make sure that obstacles do not create waste for the team. When the team needs resources—say, involvement of a potential onsite customer—the CIO helps them obtain the needed resources.

→ **Fifth, the CIO is the developer and cultivator of high-performing teams.** She hires and trains teams, acquires tools for the teams, and helps convey notions of business value to the teams. She acts as a facilitator, framing questions and encouraging decisions to be made. The strength and happiness of the DevOps teams are a critical responsibility of the CIO.

→ **Sixth, the CIO is the guardian of the IT rules, the one responsible for making sure that they don't petrify.** As we discussed in the chapter on bureaucracy, best practices, or at least practices known to work, often become encapsulated in rules; rules are a form of institutional memory. The CIO must provide incentives for experimentation and learning that can inform rules and policies.

A few words on command-and-control versus—shall we say—gentle shaping of behavior. The goal is to get the CAS to evolve in the desired direction. Heavy-handed command-and-control will generally not produce the best results, because it will not harness the creativity and diversity of the team. As a *strategic* approach, it fails. However, this does not mean that command-and-control *tactics* should never be used. Based on observations of how the teams are performing, management needs to decide on the appropriate tactical forces to apply.

For example, I have worked with software development teams that didn't want to write automated tests, because, they said, it would slow down the development process. Perhaps an orthodox Agile view would have had me facilitate their discovery that they actually are slowed down more by poor quality and rework. I didn't think it would be productive to wait for them to discover the value of testing through retrospecting and experimentation, though: I told them they had to write tests. (Well, yes, I said it with more finesse and explanation than that.)

There is a connection between this and my point about bureaucracy. As CIO, I could establish rules to guide the behavior of the teams. This is a sort of command-and-control. The rules make sense because I know they establish practices known to be good. They correspond to the pieces of Agile bureaucracy I cited in chapter 4: "If you break the build, you fix it" or "Good technical practices are not optional."

Here's the trick. The rule is designed to accomplish a business need: in this case, to maintain high quality in a way that reduces lead time (thus the automation). If the team reflects on this and finds a better way to accomplish the same objectives, then I should allow the rule to be changed. There is my CIO role in a nutshell: establish the goal, learn from the team, and render their ideas in rules that apply to future teams as a form of institutional memory. There must be a feedback cycle from the teams' learning to leadership's context-setting and rule-making.

I hope you have enjoyed this excursion through the looking glass into the land of business value. It is a world that seems familiar but where the rules might be a little different than what we expect. I've tried my best to disorient you and lead you to question some of the things we take for granted in the Agile world.

Here is the point I hope you'll take away. Business value isn't something that is well understood by *the business*. To quote Tweedledee, who appears to be have some insight into business value, "Contrariwise, if it was so, it might be; and if it were so, it would be; but as it isn't, it ain't. That's logic." Business value isn't something that is a given or that Agile teams can't question. There isn't some kind of MBA magic to it. It is something that we can all talk about and ask questions about, something we can debate and contribute to. It is something that we *must* talk about, because we can't just deliver features and expect business value to pop out of them.

There is a legacy way of thinking that distinguishes between *the business* and the technologists. The business figures out what is valuable, puts it into a set of requirements, and tosses it over the wall to IT. IT then makes a commitment to cost and schedule and delivers. This made sense in a Waterfall world. But have we really changed this way of thinking as we have moved into the new paradigms of Agile, Lean, Lean Startup, DevOps, and so on? Or have we retained elements of this legacy model by letting the product owner, onsite customer, or other agents of the business make value decisions and then charging the Agile team with "committing" each iteration to turn those business needs into solutions? Perhaps we have kept the model fundamentally the same but made it iterative and increased the amount of communication.

Changing the model would require that the development teams, and indeed the entire IT organization, be partners in deciding what is valuable. And that in turn would require that the development teams understand what business value means and be able to make decisions based on that understanding. Our journey in this book has been about demystifying business value. If there is no hard line between development and the business—if, in fact, development is part of the business—well, then, go forth and produce business value.

But understand that there is no magic formula for business value that only the business people know. Ultimately, it turns out that business value is what the business values, and that is that.

1 We can't tell the difference between one VM (or cow) and another; if we tear
   down a VM and replace it with another, we shouldn't be able to notice any change.
   Replacing Fido by Rover, on the other hand, might disturb the kiddies.

2 The Spine Model website at http://spine.wiki does not currently have very thorough
   explanation of the model. My explanation here comes from Trethewey and Roux's
   talk at Agile 2015.

3 Gojko Adzic, *Impact Mapping*, (Surrey, UK: Provoking Thoughts, 2012).

4 I owe this idea to Jeff Nielsen, in private email correspondence.

5 Ross, 47.

6 Ibid., 2.

7 Ibid., 43.

8 Ibid., 13.

9 Ibid., 201

10 If you happen to be a fan of the board game Go, I am thinking of latent possibility
   here in the sense of *aji*, one of the game's strategic concepts. Aji literally means "taste,"
   and it refers to board positions that may have value later in the game, if the player is
   able to unlock that value.

11 Douglas Hubbard, *How to Measure Anything: Finding the Value of Intangibles in Business*.
   3rd ed. (Hoboken: Wiley, 2014), pp. 145–172.

12 Humble, Molesky, and O'Reilly, p. 179, citing Ronny Kohavi et al., "Online Experimentation
   at Microsoft," http://stanford.io/130uW6X, 2009, p. 7. Kohavi et al. also cite Moran,
   "Do it Wrong Quickly: How the Web Changes the Old Marketing Rules," 2007, p. 240,
   saying that Netflix believes 90 percent of its ideas are wrong. The important point
   is that our projections about the business value we believe we will gain from a feature
   are very likely to be wrong.

13 Bjarte Bogsnes, *Implementing Beyond Budgeting: Unlocking the Performance Potential*,
   (Hoboken: John Wiley and Sons, 2009).

14 Daniel Pink, *To Sell is Human: The Surprising Truth About Moving Others* (New York:
   Riverhead Books, 2012), 21.

15 Ross, 201.

BIBLIOGRAPHY

Adler, Paul S. "The 'Learning Bureaucracy': New United Motor Manufacturing, Inc." School of Business Administration, University of Southern California. DRAFT 3.1 April 1992 (Forthcoming in Barry M. Staw and Larry L. Cummings (eds.) Research in Organizational Behavior, Greenwich, CT: JAI Press). Journal Article.

Adzic, Gojko. *Impact Mapping*. Surrey, UK: Provoking Thoughts, 2012. Kindle e-book.

Ancona, Deborah and Henrik Bresman. *X-Teams: How to Build Teams that Lead, Innovate, and Succeed*. Boston: Harvard Business School Press, 2007. Kindle e-book.

Anderson, Philip. "Seven Levers for Guiding the Evolving Enterprise" in Clippinger. pp. 113–152.

Avery, Christopher M. "Responsible Change." *Cutter Consortium Agile Project Management Executive Report* 2005 6 (10): 1–28. Journal Article.

Beck, Kent with Cynthia Andres. *Extreme Programming Explained: Embrace Change*. 2nd ed. Boston: Addison-Wesley, 2005. Book.

Bogsnes, Bjarte. *Implementing Beyond Budgeting: Unlocking the Performance Potential*. Hoboken: John Wiley and Sons, 2009. Book.

Brealey, Richard A. and Stewart C. Myers. *Principles of Corporate Finance*. 4th ed. New York: McGraw-Hill, 1991. Book.

van Cauwenberghe, Pascal. "How do you estimate the business value of user stories? You don't." December 30, 2009. http://blog.nayima.be/2009/12/. Blog Post.

Clark, Andy. "Leadership and Influence: The Manager as Coach, Nanny, and Artificial DNA" in Clippinger. pp. 47–66.

Clippinger III, John Henry ed. *The Biology of Business: Decoding the Natural Laws of Enterprise*. San Francisco: Jossey-Bass Publishers, 1999. Book.

Cohn, Mike. "The Problems with Estimating Business Value." *Mountain Goat Software*. September 30, 2010. http://www.mountaingoatsoftware.com/blog/the-problems-with-estimating-business-value. Blog Post.

——— "Topics in Scrum: Product Owner." Web page at https://www.mountaingoatsoftware.com/agile/scrum/product-owner.

——— *Succeeding with Agile: Software Development Using Scrum*. Boston: Addison-Wesley, 2010. Book.

Copeland, Tom, Tim Coller, and Jack Murrain. *Valuation: Measuring and Managing the Value of Companies*. University ed. New York: John Wiley and Sons, 1996. Book.

Dixit, Avinash K. and Robert S. Pindyck. "The Options Approach to Capital Investment." *Harvard Business Review*. May–June 1995. Journal Article.

Fitzgerald, Donna. "Principle-Centered Agile Portfolio Management." *Cutter Consortium Agile Project Management Executive Report* 6 (5). Journal Article.

Graeber, David. "In Regulation Nation" in *Harper's Magazine* March 2015. pp. 11–16. Article.

Hackman, J. Richard. *Leading Teams: Setting the Stage for Great Performances*. Boston: Harvard Business School Press, 2002. Book.

Hansen, Morten T. "IDEO CEO Tim Brown: T-Shaped Stars: The Backbone of IDEO's Collaborative Culture." ChiefExecutive.Net. http://web.archive.org/web/20110329003842/ http://www.chiefexecutive.net/ME2/dirmod.asp?sid=&nm=&type=Publishing&mod= Publications::Article&mid=8F3A7027421841978F18BE895F87F791&tier=4&id= F42A23CB49174C5E9426C43CB0A0BC46.

Highsmith, Jim. *Agile Software Development Ecosystems*. Upper Saddle River, NJ: Addison-Wesley, 2002. Book.

Highsmith, Jim et al. "Declaration of Interdependence." February 17, 2005. http://pmdoi.org. Web Page.

Horngren, Charles T., George Foster, and Srikant M. Datar. *Cost Accounting: A Managerial Emphasis*. Englewood Cliffs, NJ: Prentice-Hall, 1994. Book.

Hubbard, Douglas W. *How to Measure Anything: Finding the Value of Intangibles in Business*. 3rd ed. Hoboken: Wiley, 2014. Book.

Humble, Jez, Joanne Molesky, and Barry O'Reilly. *Lean Enterprise: How High Performance Organizations Innovate at Scale*. Sebastopol, CA: O'Reilly Media, 2015. Book.

Hunter, Richard and George Westerman. *The Real Business of IT: How CIOs Create and Communicate Business Value*. Boston: Harvard Business Press, 2009. Book.

Katz, Daniel and R. L. Kahn. *The Social Psychology of Organizations*, New York: Wiley, 1966. Book.

Kotler, Philip. *Marketing Management: Analysis, Planning, Implementation, and Control*. 8th ed. Englewood Cliffs, NJ: Prentice-Hall, 1994. Book.

Kruchten, Philippe. "The Elephants in the Agile Room." February 13, 2011. http://philippe. kruchten.com/2011/02/13/the-elephants-in-the-agile-room/. Blog Post.

Larman, Craig and Bas Vodde. *Scaling Lean and Agile Development: Thinking and Organizational Tools for Large-Scale Scrum*. Upper Saddle River, NJ: Addison-Wesley, 2009. Book.

Leffingwell, Dean. *Agile Software Requirements: Lean Requirements Practices for Teams, Programs, and the Enterprise*. Boston: Addison-Wesley, 2011. Book.

Maizlish, Bryan and Robert Handler. *IT Portfolio Management Step by Step: Unlocking the Business Value of Technology*. Hoboken: John Wiley and Sons, 2005. Book.

"Manifesto for Agile Development," February 11–13, 2001. Web page at http://agilemanifesto.org.

Mill, John Stuart. *Considerations on Representative Government*. 2nd ed. London: Parker, Son, and Bourn, 1865. Google Book.

Moore, Mark H. *Creating Public Value: Strategic Management in Government*. Cambridge, MA: Harvard University Press, 1995. Book.

Moreira, Mario E. *Being Agile: Your Roadmap to Successful Adoption of Agile*. New York: Apress, 2013. Book.

Pichler, Roman, "Being an Effective Product Owner." Web page at https://www.scrumalliance.org/community/articles/2007/april/being-an-effective-product-owner.

Pink, Daniel H. *To Sell is Human: The Surprising Truth About Moving Others*. (New York: Riverhead Books, 2012). Kindle e-book.

Poppendieck, Mary and Tom Poppendieck. *Lean Software Development: An Agile Toolkit*. Upper Saddle River, NJ: Addison-Wesley, 2003. Book.

Porter, Michael. *Competitive Strategy: Techniques for Analyzing Industries and Competitors*. New York: Free Press, 1985. Book.

Rappaport, Alfred. *Creating Shareholder Value: A Guide for Managers and Investors*. New York: The Free Press, 1998. Book.

Reinertsen, Donald G. *The Principles of Product Development Flow: Second Generation Lean Product Development*. Redondo Beach, CA: Celeritous Publishing, 2009. Book.

Ries, Eric. *The Lean Startup: How Today's Entrepreneurs Use Continuous Innovation to Create Radically Successful Businesses*. New York: Crown Business, 2011. Book.

Ross, Jeanne W., Peter Weill, and David C. Robertson. *Enterprise Architecture as Strategy: Creating a Foundation for Business Execution*. (Boston: Harvard Business School Press, 2006). Book.

Schwaber, Ken and Jeff Sutherland. "The Scrum Guide: The Definitive Guide to Scrum: The Rules of the Game" PDF at Scrum.org. October 2011.

Schwaber, Ken. *Agile Project Management with Scrum*. Redmond, WA: Microsoft Press, 2009. Kindle e-book.

Schwartz, Peter. *The Art of the Long View: Planning for the Future in an Uncertain World*. New York: Currency Doubleday, 1991. Book.

Schein, Edgar H. *The Corporate Culture Survival Guide*. Revised ed. San Francisco: Jossey-Bass, 2009. Book.

Smith, Greg and Ahmed Sidky. *Becoming Agile in an Imperfect World*. Greenwich, CT: Manning Publications, 2009. Book.

Sutherland, Jeff . "Ten Year Agile Retrospective: How We Can Improve in the Next Ten Years." Microsoft Developer Network (MSDN), https://msdn.microsoft.com/en-us/library/hh350860(v=vs.100).aspx.

Takeuchi, Hirotaka and Ikujiro Nonaka. "The New New Product Development Game" in *Harvard Business Review* Jan–Feb 1986. Journal Article.

Watson, Jim, Michael Rosen, and Kurt Guenther. "Are Agile Methods and Enterprise Architecture Compatible? Yes, With Effort." *Cutter Consortium Agile Project Management Executive Report* 6 (11). Journal Article.

Weber, Max ed. by Guenther Roth and Claus Wittich. *Economy and Society: An Outline of Interpretive Sociology*. Berkeley: University of California Press, 1978. (Originally published 1922.) Book.

Weill, Peter and Jeanne W. Ross. *IT Governance: How Top Performers Manage IT Decision Rights for Superior Results*. Boston: Harvard Business School Press, 2004. Book.

Wren, Daniel and Arthur Bedeian. *The Evolution of Management Thought*. 6th ed. Hoboken: John Wiley and Sons, 2009. Book.

Zietlow, John, Jo Ann Hankin, and Alan Seidner. *Financial Management for Nonprofit Organizations: Policies and Practice*. Hoboken: Wiley, 2007. Book.

ACKNOWLEDGMENTS

The DevOps community—and by extension today's Enterprise IT community—has an animating spirit: Gene Kim. Gene's support has been tremendous, surprising, deeply impactful, and absolutely critical in assembling the ideas in this book.

I was extremely lucky to have an awe-inspiring group of reviewers help me with their feedback. Jez Humble, Adrian Cockcroft, Gary Gruver, Larry Goldberg, Jeff Nielsen, Matt Rubins, Eric Willeke, Erez Tatcher, and of course, Gene Kim. My editor Robyn Crummer-Olson and publisher Todd Sattersten said just the right things at the right moments to spark the ideas that made their way into the book. They also got me to stop tweaking sections that were already okay—you know how software developers are. Thank you all so much.

Writing this book was a way for me to organize and reflect on the things that I have learned as a CIO and CEO. I had many partners in learning these lessons, and they played an important role in it. I'd like to thank my entire team at USCIS, and especially Keith Jones, Kath Stanley, and Joshua Seckel, who fought all the battles alongside me. Special thanks also to Rendell Jones and Tracy Renaud, my managers, who helped me understand the logic of the government. Take Yokota and John Wilhelm might be surprised to find themselves starring in a few anecdotes in this book; I don't know if they realized how much I learned working with them and the rest of the talented, passionate people at Intrax: Heidi Mispagel, Sherry Carpenter, Paul Bydalek, and everyone else.

There is a big change going on across federal government IT now, and it is reflected in the ideas in this book. Luke McCormack and Margie Graves cleared the obstacles to allow me to experiment, while our team from US Digital Service (Brian, Will, Eric, Grant, Nacin, Dana, Nick, Victor, Aalok, and Mollie) have taught me new tricks and forced me always to be rigorous in my thinking.

Oh, and thanks to Jenny, who is convinced that my butt has become glued to my writing chair.

ABOUT THE AUTHOR

Mark Schwartz is an iconoclastic CIO and a playful crafter of ideas, an inveterate purveyor of lucubratory prose. He has been an IT leader in organizations small and large, public, private, and nonprofit. As the CIO of US Citizenship and Immigration Services, he provokes the federal government into adopting Agile and DevOps practices. He is pretty sure that when he was the CIO of Intrax Cultural Exchange, he was the first person ever to use business intelligence and supply chain analytics to place au pairs with the right host families. Mark speaks frequently on innovation, bureaucratic implications of DevOps, and Agile processes in low-trust environments. With a computer science degree from Yale and an MBA from Wharton, Mark is either an expert on the business value of IT or just confused and much poorer.